**SI calculations**

**in engineering science**

# SI calculations in engineering science

**A. C. Walshaw**

*Professor Emeritus*

*Formerly Dean of the Faculty of Engineering,*
*The University of Aston in Birmingham*

**NEWNES–BUTTERWORTHS**
LONDON BOSTON
Sydney Wellington Durban Toronto

The Butterworth Group

| | |
|---|---|
| United Kingdom | **Butterworth & Co. (Publishers) Ltd.**<br>London: 88 Kingsway, WC2B 6AB |
| Australia | **Butterworths Pty Ltd.**<br>Sydney: 586 Pacific Highway, Chatswood NSW 2067<br>Also at Melbourne, Brisbane, Adelaide and Perth |
| Canada | **Butterworth & Co. (Canada) Ltd.**<br>Toronto: 2265 Midland Avenue, Scarborough, Ontario M1P 4S1 |
| New Zealand | **Butterworths of New Zealand Ltd.**<br>Wellington: 26–28 Waring Taylor Street, 1 |
| South Africa | **Butterworth & Co (South Africa) (Pty) Ltd.**<br>Durban: 152–154 Gale Street |
| USA | **Butterworth (Publishers) Inc.**<br>Boston: 19 Cummings Park, Woburn, Mass. 01801 |

First published 1977



ISBN 0 408 00284 0

Typeset by Reproduction Drawings Ltd,
Sutton, Surrey

Printed in England by The Whitefriars Press Ltd,
London and Tonbridge

# Preface

This book gives a concise and straightforward presentation of Le Système International d'Unités (the SI) and illustrates its use in a selection of worked examples involving mechanics, fluids, thermodynamics and graphs. Its main aim is to help students, practising engineers and designers develop a facility for using and manipulating units with assurance, and dealing confidently with quantitative work generally.

The letter symbols used for physical quantities and units are those recommended by the international standards organisations, namely, the Conférence Général des Poids et Mesures (CGPM) and the International Organisation for Standardisation (ISO).

Special attention is given to the derivation and use of physical and numerical (including empirical) formulae. Although the former are the most important in that they are universal and independent of particular units or systems of units, students and engineers encounter and have to use both types. Hence, *it is essential that there be no doubt as to how all formulae are derived before they are used*. Also, students of science and technology should not only be 'dimension conscious', but have a sure method of manipulating, checking and reducing units in calculations. Hence, the unity multipliers or 'unity brackets' used in a range of typical examples which have been worked out in the text, and which readers may first try to solve for themselves.

A.C.W.

Aston in
Birmingham 1977

# Contents

1

# Symbols for physical quantities and units

Just as there is a distinction between physical quantities (e.g. mass, $m$; length, $l$; time, $t$) and units (e.g. kilogram, kg; metre, m; second, s), a distinction is made between their respective symbols; and clearly units differ in magnitude (e.g. second and hour) and in kind (e.g. second and metre).

There are insufficient letters to enable each to be allocated only to a single quantity, but there are sufficient capital and lower-case letters and subscripts to ensure that no ambiguity need arise in the symbols used in any particular branch of science or technology (see alphabetical list on pages 2–4).

## Symbols for physical quantities

These are usually single letters of Latin or Greek alphabets printed in italic (*sloping*) type, although two-letter symbols (from proper names) are used to represent dimensionless groups of certain physical quantities: e.g Mach number, $Ma = u/a$ (see Example 20); Reynolds number, $Re = ul/v$. Such two-letter symbols can have their indivisibility stressed, if necessary, by placing them within brackets ($Re$), ($Pr$), ($Nu$) (see Examples 21 and 26).

## Symbols for units

These are printed in a lower-case roman (upright) type, except where a unit is derived from a proper name from which a capital roman letter is taken as the unit symbol (e.g. A for the ampere, K for the kelvin, N for the newton, J for the joule, Pa for pascal = $\mathrm{N/m^2}$). Symbols for chemical elements are also printed in capital (upright) letters (e.g. $\mathrm{C + O_2 = CO_2}$).

Unit symbols do not take an 's' in plurals (e.g. that for moles is mol, not mols, and kilograms is kg, not kgs); and in division of one unit by others, only one solidus (/) is used (e.g. acceleration in metres per second per second is written m/s$^2$, not m/s/s, and thermal conductivity is J/s K m, not J/s/K/m). For example, if the temperature-difference of two sides of a flat metal plate 8.0 mm thick is 10 K and the coefficient of conductivity is 240 MJ mm/m$^2$ h K or $\frac{1}{15}$ kW/m K, the rate of heat-transfer is given by

$$q = \frac{Q}{t} = k\frac{(T_1 - T_2)A}{d}$$

i.e.
$$\frac{q}{A} = k\frac{(T_1 - T_2)}{d} = \left(240\,\frac{\text{MJ mm}}{\text{m}^2\text{ h K}}\right)\frac{10\text{ K}}{8\text{ mm}}$$

$$= 300\text{ MJ/m}^2\text{ h}$$

or alternatively,
$$\frac{q}{A} = \left(\frac{1\text{ kW}}{15\text{ m K}}\right)\frac{10\text{ K}}{0.008\text{ m}} = \frac{1}{12}\text{ MW/m}^2$$

**In hand-writing and typescript**

The distinction can be made when necessary by underlining the symbols for physical quantities.

**Alphabetical lists for reference**

Letter symbols and unit symbols used in engineering science are listed here for easy reference in case there is doubt when studying the worked examples.

*Letter symbols for physical quantities*

| | |
|---|---|
| $A$ | area |
| $a$ | area, acceleration, velocity of sound in gas |
| $B$ | buoyancy force |
| $b$ | breadth |
| $C$, $C_m$ | heat-capacity, molar heat-capacity, coefficient of drag ($C_D$) |
| $c$ | specific heat-capacity ($C/m$) |
| $c_v$, $c_p$ | specific heat-capacity at constant volume/pressure |
| $d$ | diameter, distance, relative density ($\rho/\rho_{(H_2O)}$) |

| | |
|---|---|
| $E$ | energy, Young's modulus, internal energy |
| $e$ | specific internal energy ($E/m$) |
| $F$ | force |
| $f$ | cyclic frequency, function of |
| $G$ | modulus of rigidity |
| $Gr$ | Grashof number ($\beta\rho^2 l^3 g\Delta T/\eta^2$) |
| $g$ | acceleration of free fall |
| $H, H_m$ | enthalpy ($E + PV$), molar enthalpy ($H/n$), head of fluid |
| $h$ | coefficient of heat-transfer, specific enthalpy ($H/m$) |
| $I$ | electric current, moment of inertia, second moment of area |
| $J$ | second polar moment of area |
| $k$ | thermal conductivity, elastic or spring stiffness |
| $L$ | length |
| $l$ | length |
| $M$ | bending moment, molar mass ($m/n$) |
| $M_r$ | relative molecular mass (formerly 'molecular weight') |
| $Ma$ | Mach number ($u/a$) |
| $m, \dot{m}$ | mass, rate of mass flow |
| $N, n$ | rotational speed, rotational frequency |
| $n$ | amount of substance (number of moles) |
| $Nu$ | Nusselt number ($hD/k$) |
| $P$ | power, pressure |
| $Pr$ | Prandtl number ($c_p\eta/k$) |
| $p$ | pressure |
| $Q, \dot{Q}$ | quantity of heat-transfer, volumetric rate of flow of fluid |
| $q$ | rate of heat-transfer |
| $R$ | specific gas constant, electric resistance, radius |
| $\bar{R}$ | molar (or universal) gas constant ($Rm/n$ or $RM$) |
| $Re$ | Reynolds number ($ul/\nu$) |
| $r$ | radius, compression ratio ($r_v$), pressure ratio ($r_p$) |
| $S, S_m$ | entropy, molar entropy ($S/n$), distance |
| $s$ | distance, specific entropy ($S/m$), scales of graphs |
| $T$ | thermodynamic temperature (kelvin), torque, thrust force |
| $t$ | time, customary temperature (Celsius) |
| $U$ | velocity, overall heat-transfer coefficient |
| $u$ | velocity |
| $V, V_m$ | volume, electric pressure-difference (volt), molar volume ($V/n$) |
| $v$ | specific volume ($V/m$), velocity |
| $W$ | quantity of work, weight ($mg$) |
| $w$ | specific work ($W/m$), load per unit length of beam |
| $X, x$; $Y, y$ | distance, Cartesian coordinates |
| $Z, z$ | section modulus, height above a datum |
| $\alpha$ | plane angle, angular acceleration |
| $\gamma$ | ratio $c_p/c_v$, shear strain or shear angle ($\Delta\theta/\theta_0$) |

| | |
|---|---|
| $\delta$ | difference, increment, deflection of beams |
| $\epsilon$ | emissivity, linear strain |
| $\eta$ | dynamic viscosity, efficiency |
| $\theta$ | plane angle |
| $\lambda$ | wavelength |
| $\mu$ | coefficient of friction |
| $\nu$ | kinematic viscosity ($\eta/\rho$) |
| $\pi$ | ratio circumference/diameter of circle (3.1416. . . ) |
| $\rho$ | density ($m/V = 1/v$) |
| $\tau$ | periodic time, shear stress |
| $\phi$ | angle, function of |
| $\omega$ | angular velocity or angular frequency |

*Unit symbols and special names*

| *Base units of the SI (see also Chapter 2)* | |
|---|---|
| A | ampere |
| cd | candela |
| K | kelvin |
| kg | kilogram |
| m | metre |
| mol | mole |
| s | second |

| *Supplementary SI units* | |
|---|---|
| rad | radian |
| sr | steradian |

| *Derived units and special names (see also Chapter 2)* | |
|---|---|
| Hz | hertz |
| J | joule |
| l | litre |
| lm | lumen |
| lx | lux |
| N | newton |
| P | poise |
| Pa | pascal |
| St | stokes |
| t | tonne |
| W | watt |
| V | volt |
| Ω | ohm |

| *Internationally agreed deviations from the SI* | |
|---|---|
| h | hour |
| kW h | kilowatt-hour |
| min | minute |
| ($\pi$/180) rad | degree (angle) |
| °C | degree Celsius |

*Note* So as not to confuse the unit symbol (l) for litre with the figure one (1), the word litre is often written in full.

### Coherence of the SI

Although measures and systems of units are arbitrary as devised by man, whereas laws of nature interconnecting physical quantities (e.g. $F = ma$ and $V = IR$) are not, the need for a unified system or universal frame of reference and international language so far as units, symbols and abbreviations are concerned, has long been recognised and centred on the Conférence Général des Poids et Mesures (CGPM) and the International Organisation for Standardisation (ISO).

Each country, however, has its Standards Bureau or Institution for coordinating the views of industry and professional bodies, and for disseminating information and recommendations on units, symbols and abbreviations, and standards generally.

In addition to the major advantage of having international standards facilitating the exchange of scientific, industrial and commercial information, time and effort is saved in teaching and learning, and risks of errors minimised in calculations, as a result of adopting a metric system of which the SI is the favoured version resulting from international considerations.

Previous systems of units were not coherent, because they involved many awkward numerical factors (e.g. 1 lb = 16 oz; 1 mile = 1760 yd; 1 acre = 4840 $yd^2$, etc.) as compared with metric systems, which merely involve factors of ten.

In a coherent system all derived units are obtained by multiplication or division of the base units without the introduction of any numerical factors (even powers of ten); for example, the SI unit of force derived from Newton's second law of motion ($F = ma$) is the newton, $N = kg\,m/s^2$ (which is independent of gravity–'standard' or otherwise); and the SI derived unit of power is the watt, W = N m/s = J/s.

### Example 1

The pressure ($p$) at a depth ($d$) of 100 m in water of density ($\rho$) $10^3$ kg/m$^3$ or 1 Mg/m$^3$ or 1 t/m$^3$ is

$$p = (\rho g)d$$

$$= 10^3 \frac{\text{kg}}{\text{m}^3} \times 9.807 \frac{\text{m}}{\text{s}^2} \times 100\ \text{m} \left[\frac{\text{N s}^2}{\text{kg m}}\right]$$

$$= 980.7\ \text{kN/m}^2 \text{ (gauge, i.e. above atmospheric)}$$

and the depth at which the pressure is 100 kN/m$^2$ above that of the atmosphere is

$$d = \frac{p}{\rho g} = 100 \frac{\text{kN}}{\text{m}^2} \times \frac{\text{m}^3}{9.807\ \text{kN}} = 10.2\ \text{m}$$

**Example 2**

(a) Calculate the kinetic energy of an automobile of mass 1000 kg when travelling with a speed of 100 km/h. Also estimate the force which would be exerted in a crash bringing the car to rest (assuming constant retardation) in 2.78 s.

$$\text{Kinetic energy } E = \tfrac{1}{2} mu^2 = \tfrac{1}{2} \times 1000\ \text{kg} \times \left(\frac{100\ \text{km}}{3600\ \text{s}}\right)^2$$

$$= 500 \left(\frac{100}{3.6}\right)^2 \text{kg} \frac{\text{m}^2}{\text{s}^2} \left[\frac{\text{N s}^2}{\text{kg m}}\right]$$

$$= 386\,000\ \text{N m or } 386\ \text{kJ}$$

$$\text{Impact force } F = ma = 1000\ \text{kg} \times \frac{100\,000\ \text{m}}{3600\text{s} \times 2.78\ \text{s}} \left[\frac{\text{N s}^2}{\text{kg m}}\right]$$

$$= 10\,000\ \text{N or } 10\ \text{kN}$$

(b) A bullet of mass 50 g fired into a sandbag of mass 50 kg suspended on a rod 1 m long caused the straight rod to deflect from its vertical position by an angle of 14°6′. Neglecting the mass of the rod, estimate the velocity and kinetic energy of the bullet on striking the bag.

Applying two fundamental laws of mechanics, namely (i) conservation of energy and (ii) conservation of momentum, we have:

(i) if $u_1$ is the velocity of the bullet before impact, and $u_2$ is the velocity of the bullet plus bag after impact, then, since $\frac{1}{2} Mu_2^2 = Mgh$, where $M = 50$ g + 50 kg and $h = (1 - \cos 14°6')$ m $= (1 - 0.970)$ m = 30 mm,

$$u_2 = \sqrt{(2\,gh)} = \sqrt{(2 \times 9807 \frac{\text{mm}}{\text{s}^2} \times 30\ \text{mm})} = 766\ \text{mm/s}$$

(ii) equating the momentum before and after impact,

$$50\text{g} \times u_1 + 0 = (50 + 50 \times 10^3)\,\text{g} \times 0.766\ \text{m/s}$$

i.e. $u_1 = (1 + 10^3) \times 0.766\ \text{m/s} = 766\ \text{m/s}$

the velocity of the bullet on striking the bag.
The kinetic energy of the bullet is

$$\tfrac{1}{2} m u_1^2 = \tfrac{1}{2} \times 50\ \text{g} \times 588 \times 10^3 \frac{\text{m}^2}{\text{s}^2} \left[\frac{\text{N s}^2}{\text{kg m}}\right]$$

$$= 14.7 \times 10^3\ \text{N m} = 14.7\ \text{kJ}$$

**Advantages of the SI**

The SI eliminates any confusion between mass and weight, since, being different in kind, their units have now been given different names, namely, kilogram and newton–e.g. on the earth a mass of 1 kg weighs 9.807 N.

*There is one, and only one, unit in the SI for each kind of quantity.* For example *the joule*, J = N m = kg m$^2$/s$^2$, is the derived unit of *energy* whether it be internal, potential, or in transition as work, heat-transfer or electrical energy (but see Note on page 18). Similarly, power or rate of transmission of energy (mechanical, hydraulic, electrical etc.) can equally well be expressed in terms of one derived unit of the SI, namely, *the watt*, W = J/s. Also, the numbers in physical constants are rendered universal by the SI, e.g. the speed of light in empty space is $c$ = 299.7925 Mm/s, the universal gas constant $\bar{R}$ = 8.3143 J/mol K, the Stefan-Boltzmann radiation constant $\sigma$ = 56.697 nW/m$^2$ K$^4$ used in the formula $E = \epsilon\sigma (T_1^4 - T_2^4)$ in estimating the rate at which energy radiates from a 'grey' body of emissivity $\epsilon$, and Wien's radiation (displacement) law is $\lambda_{max} T = 2.8978 \times 10^{-3}$ m K.

It will become clear that advantages of the SI are:

(a) uniformity, in that it provides a consistent and coherent system or 'universal language' so far as units are concerned;
(b) there is one, and only one, base unit for each physical quantity;
(c) coherency, in that the product or quotient of base units is the unit of the resulting quantity, i.e. derived units are obtained without introducing any numerical factors (even powers of ten);
(d) appropriate sizes (decimal fractions and multiples) of units required in different branches of science, industry and commerce can be provided by use of internationally agreed prefixes signifying multiples and submultiples of 10 (see page 13);
(e) numbers in physical constants are rendered universal.

**Example 3**

A spring balance has a scale graduated in newtons (N) and records the weight of a package as 12 N at a place on the earth where the acceleration of free fall is the so-called 'standard', namely $g_n = 9.80665 \text{ m/s}^2$. What would the balance record when weighing the same package on the moon where $g$ is one-sixth that on the earth, and what is the specific force of gravity on the earth?

$F = ma$ and, in particular, weight $W = mg$. Also, by definition of N, we deduce that

$$\left[\frac{\text{N s}^2}{\text{kg m}}\right]$$

is unity [1] and may, therefore, be used as a 'unity multiplier' (or 'newtonian unity bracket') whenever a mass/force ratio needs to be reduced by cancellation of like units, as in Examples 1 and 2.

Hence, the mass of the package (as calculated from information obtained on the earth) is

$$m = \frac{W_n}{g_n} = \frac{12 \text{ N}}{9.80665 \text{ m/s}^2}\left[\frac{\text{kg m}}{\text{N s}^2}\right] = 1.22 \text{ kg}$$

which *mass* remains the same wherever the package is situated. Thus on the moon the same mass would weigh

$$W = mg = m\frac{g_n}{6}$$

and the pointer on the scale would record for a *mass* of 1.22 kg a *weight* of 2.0 N as compared with 12 N on the earth.

Specific force is force per unit mass (see also page 18). Hence, the specific force of gravity on the earth's surface is

$$\frac{W_n}{m} = g_n \approx 9.81 \text{ m/s}^2 \approx 9.81 \text{ N/kg}$$

**Example 4**

Using Newton's second law of motion, calculate the velocity, $u_c$, which a rocket, irrespective of its mass, must acquire in order to escape from the gravitational attraction of the earth and obey Newton's first law of motion. Neglect the loss of mass due to burning of fuel, and assume that

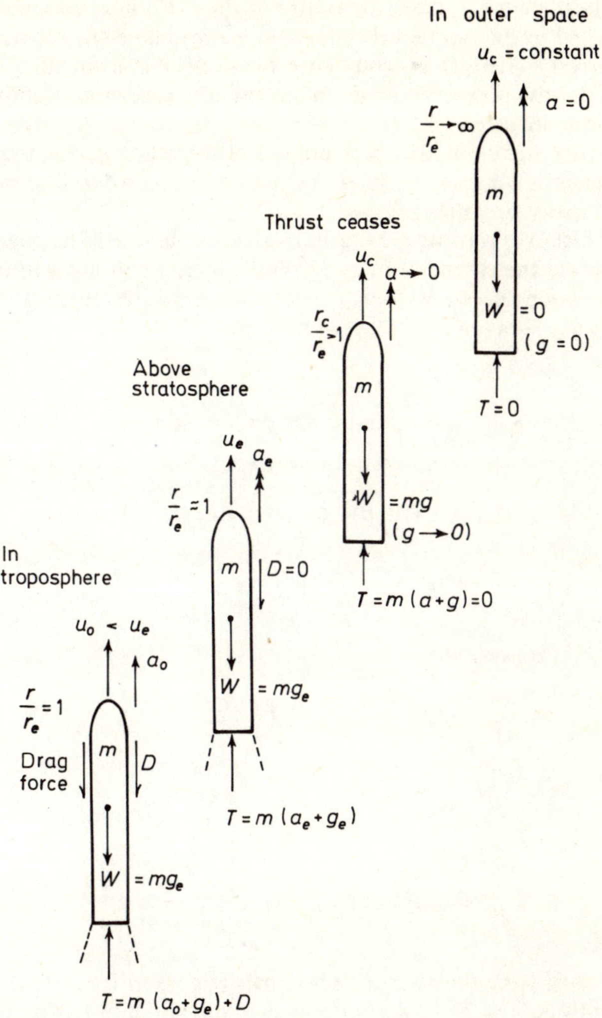


*Figure 1.1 Rocket travelling from earth to outer space*

the acceleration of free fall, $g$, varies as the square of the distance from the centre of the earth, the radius of which is $r_e = 6.45$ Mm.

Referring to *Figure 1.1* we see that:

(a) the resultant force on the rocket just after lift-off is $F = T - W - D = ma_o$ and hence the thrust of the tailjet is $T = m\,(a_o + g_e) + D$, where the drag force, $D$, is proportional to $u_o^2$.

(b) as the distance, $r$, from the centre of the earth increases, a point is reached in the rarefied air above the stratosphere ($r/r_e$ remaining nearly 1) where $D = 0$ and hence $T = m\,(a_e + g_e)$, which enables the powerful thrust to increase the acceleration more rapidly until

(c) the fuel burns out, at which point $T = 0 = m\,(a + g)$, i.e. the acceleration $a = -g = -g_e\,(r_e/r)^2$ which, as $r$ continues to increase, eventually becomes zero as

(d) the rocket leaves the gravitational field of the earth, having achieved the escape velocity $u_c$ with which it continues in outer space—weightless, without acceleration and without thrust, i.e. obeying Newton's first law of motion.

In general,

$$\text{acceleration } a = \frac{\mathrm{d}u}{\mathrm{d}t} = \frac{\mathrm{d}u}{\mathrm{d}r}\,\frac{\mathrm{d}r}{\mathrm{d}t} = u\,\frac{\mathrm{d}u}{\mathrm{d}r}$$

Hence, at the point (c) where the gas jet ceases (i.e. $T = 0$),

$$a = -g = -\left(\frac{r_e}{r}\right)^2 g_e = u\,\frac{\mathrm{d}u}{\mathrm{d}r}$$

from which it follows that

$$\int_{u_e}^{u} u\,\mathrm{d}u = -g_e r_e^2 \int_{r_e}^{r} \frac{\mathrm{d}r}{r^2}$$

and

$$\frac{u^2 - u_e^2}{2} = -g_e r_e^2\left(\frac{1}{r_e} - \frac{1}{r}\right)$$

or

$$u^2 = u_e^2 - 2g_e r_e\left(1 - \frac{r_e}{r}\right)$$

Thus, however large the distance $r$ becomes (e.g. even if $r_e/r \to 0$ and consequently, $u^2 \to u_e^2 - 2g_e r_e$) the rocket will still move away from the earth (i.e. $u$ will be positive) provided that $u_e^2 > 2g_e r_e$, i.e. the minimum escape velocity is

$$u_c = \sqrt{(2 \times 9.07\ \text{m} \times 6.45 \times 10^6\ \text{m})} = 11.25\ \text{km/s}$$

or

$$11.25\,\frac{\text{km}}{\text{s}}\left[\frac{\text{s knot}}{0.514\ \text{m}}\right] = 21\,900\ \text{knot}$$

We may note that the 'inertia' of a mass may be thought of as its reluctance to be accelerated (or, personally, as reluctance to shift oneself!).

2

# Base units and derived units of the SI, special names and deviations

Whatever system of units is used it is necessary for the magnitudes of a selection of basic physical quantities to be arbitrarily declared to have unit value. *The SI has seven base units.* This does not imply that base units are more fundamental than other units, but that they were chosen as convenient and acceptable units from which all other units needed in science and technology could be derived coherently, i.e. by simple processes of multiplication and division of base units with unity as the only multiplying factor involved, thus giving elegance to the SI as compared with previous non-metric systems.

**Base units of the SI (definitions are given in Appendix 1)**

| *Physical quantity* | *Name of unit* | *Unit symbol* | *Letter symbol for physical quantity* |
|---|---|---|---|
| Length | metre | m | $l$, $L$ |
| Mass | kilogram | kg | $m$ |
| Time | second | s | $t$ |
| Electric current | ampere | A | $I$ |
| Thermodynamic temperature | kelvin | K | $T$ |
| Luminous intensity | candela | cd | $I_v$ |
| Amount of substance | mole | mol | $n$ |

Base units are defined with reference to a set of standards, whereas derived units are defined in terms of base units.

The kilogram is a legal unit of mass equal to that of the international prototype kilogram, which is a cylinder of platinum alloy kept by the Bureau International des Poids et Mesures at Sèvres. The retention (because of widespread use) of the name 'kilogram' has, of course, been criticised on the ground that a base unit ought not to have a multiplying prefix (k), but a new name has not been agreed.

The metre is now defined in terms of a number of wavelengths of a particular wavelength of light. This 'optical metre' is more accurate than any other linear standard and can be reproduced independently anywhere in the world without having to refer to the international prototype metre in Sèvres, where the platinum–iridium bar (representing one-ten-millionth part of the earth's meridional quadrant) is kept.

The CGPM decided (1967) that the SI base unit kelvin (K) be used both for thermodynamic (commonly referred to as *absolute*) temperature and for temperature-interval. For those at present unfamiliar with the second law of thermodynamics (and equation $T_1/T_2 = Q_1/Q_2$), and because of common usage of the empirical Celsius scale, '°C' continues to be used as a scale-indicator or label attached to numbers indicative of temperature levels on the Celsius scale, i.e. $t = b$ Celsius will be written $b$°C, where °C is an abbreviation for *Celsius*. However, $b$°C implies that the thermodynamic temperature or so-called 'absolute temperature' $T = (b + 273.15)$ K, and the temperature-interval between $b_1$°C and $b_2$°C is $(b_1 - b_2)$ kelvins, i.e. $(b_1 - b_2)$ K (see Appendix 3). The kelvin (K) is defined by the statement that the thermodynamic temperature at the triple point of natural water (i.e. that saturation state at which ice, water and steam are together in equilibrium) is 273.16 K exactly. The normal freezing point of water in air at atmospheric pressure is 273.15 K (0°C) and the boiling point is 373.15 K (100°C)–the temperature at the triple point of water being 0.01 °C exactly.

The physical quantity termed 'amount of substance' and its magnitude could *in principle* (though hardly in practice) be determined by counting particles–the number of designated particles constituting one mole being $602.252 \times 10^{21}$! Thus, the mole is related to a count of a number of parts; and because this concept of 'amount of substance' does not depend on that of any other physical quantity, it is regarded as fundamental, and the CGPM decided (1971) that mole (whose unit symbol is mol) be a base unit. The accurate estimation of the mole has, however, to be in terms of relative molecular mass (a pure number, formerly called 'molecular weight') and a unit of mass (1 g) (see Appendix 5). Thus the mole is, *in practice*, a derived unit and may be regarded as a unit of mass, the magnitude of which varies from substance to substance. For example, the amount of substance ($n = m/M$) in a mass ($m$) of 24 g of carbon is 24/12 mol = 2 mol, and in 18 g of $H_2O$ is 1 mol, the molar masses ($M$) of C and $H_2O$ being 0.012 kg/mol and 0.018 kg/mol respectively.

**Supplementary units**

The CGPM permits the use of the following two units, supplementary to the seven formally adopted base units of the SI.

| *Quantity* | *Name of unit* | *Unit symbol* | | *Letter symbol* |
|---|---|---|---|---|
| Plane angle | radian | rad | dimensionless | $\alpha, \beta, \ldots \theta, \phi$ |
| Solid angle | steradian | sr | | $\Omega, \omega$ |

Thus the plane angle subtended by a semi-circle is $\pi$ rad. Alternatively, we may say that the radian measure of the plane angle of a semi-circle is $\pi$, arc/radius being dimensionless, as is the solid angle of a hemisphere, namely $2\pi r^2/r^2 = 2\pi$, usually written $2\pi$ sr in order to indicate the mode of measurement.

## Prefixes used in the SI

### *Decimal multiples and sub-multiples*

To overcome the difficulty that the sizes of the base units of the SI are not always convenient for certain purposes, and in order to establish a comprehensive specification for units of measurement, approved prefixes are used.

The internationally agreed names and symbols for multiples and sub-multiples of ten are as follows:

| *Fraction* | *Prefix name* | *Abbreviation* | *Multiple* | *Prefix name* | *Abbreviation* |
|---|---|---|---|---|---|
| $10^{-1}$ | deci | d | 10 | deca | da |
| $10^{-2}$ | centi | c | $10^{2}$ | hecto | h |
| $10^{-3}$ | milli | m | $10^{3}$ | kilo | k |
| $10^{-6}$ | micro | $\mu$ | $10^{6}$ | mega | M |
| $10^{-9}$ | nano | n | $10^{9}$ | giga | G |
| $10^{-12}$ | pico | p | $10^{12}$ | tera | T |

In practice, to reduce the number of prefixes which could be used, *preferred prefixes* have the form $10^{\pm 3n}$, where $n$ is an integer.

### *Prefixes in numerator and/or denominator*

Since the use of prefixes in both numerator and denominator would enable a physical quantity to be expressed in a large number of different ways (e.g. the 'bar', defined as $10^5$ N/m$^2$, is also daN/cm$^2$, dN/mm$^2$, MN/dam$^2$, etc.–all of which are 'correct', but which could cause an undesirable and unnecessary cluttering-up of the mind by the use of prefixes in deviating from the base units, one object of which is to simplify and unify rather than complicate), it is recommended that *prefixes be used in the numerator only, as a general rule.* There is then less likelihood of errors, e.g. MJ/dm$^3$ = $10^6$ J/($10^{-1}$ m)$^3$ = $10^9$ J/m$^3$ is clearer and less cumbersome if written GJ/m$^3$, as also is Young's modulus $E$ = 207 kN/mm$^2$ if written 207 GN/m$^2$. In some cases, however, pressures and stresses are conveniently expressed in N/mm$^2$, which is alternatively in MN/m$^2$.

Thus, by means of prefixes (which are printed or written *immediately adjacent to unit symbols*) a range of sizes involving reasonably sized numbers can be achieved. For example, rather than express all pressures and stresses in the principal or base units of the SI, N/m$^2$ (given the special name 'pascal' and unit symbol Pa), a range of units of stress can be provided, namely . . . μPa, mPa, Pa, kPa, MPa . . . , suitable for every purpose in science and engineering (the 'bar', being $10^5$ Pa, is 'non-preferred'). *Thus the SI has the advantage that all units of the same physical quantity are related by powers of* 10, *preferably by* $10^{\pm 3n}$.

### *Single-prefix rule*

*Only one prefix is applied at one time to a unit symbol,* i.e. compound prefixes should not be used. For example 1 microgram = $10^{-6}$ g is written 1 μg, not 1 mmg or 1 nkg, and 1 megagram is written 1 Mg, not 1 kkg. Thus, although the name 'kilogram' has been retained rather than a new one invented for the SI base unit of mass, the single-prefix rule is applied in that *kg is regarded as a prefixed unit,* e.g. the unit symbol for 'milli-kilogram' is written g, not mkg. Similarly for dynamic viscosity (which can be expressed in Pa s) 1 centipoise (1 cP) is not written 1 cdN s/m$^2$ but as 1 mN s/m$^2$ (alternatively 1 mPa s). It will be seen that *care is needed with the* 'm' *to avoid confusion between* m *used as a prefix for 'milli'* = $10^{-3}$ *and as a unit symbol for metre;* e.g. mN is a prefixed symbol meaning milli-newton and not metre × newton, for the latter should be written m × N, or m . N, or, best of all, N m, in order to eliminate any possible confusion between mN and N m. Hence prefixes are written immediately adjacent to unit symbols with which they are associated (e.g. MJ, kW), whereas multiplication of unit symbols is indicated by a gap, full point or multiplication sign (e.g, kW h, mN m, mN.m or mN × m, and it is safer to write knot s as knot × s or s knot).

### *Prefixed units raised to powers*

*The prefix is part of the unit,* i.e. the combination of a prefix and a unit symbol is *considered as one new unit symbol.* Thus, when a multiple or sub-multiple of a base SI unit is raised to a power, *the index applies* (without indicating by brackets) *to the whole prefixed unit* and not to the base unit alone. For example, 1 $km^2$ is 1 $(km)^2$ = $10^6$ $m^2$, not 1000 $m^2$; likewise $hm^2$ = $10^4$ $m^2$ = hectare, and $dm^3$ = $10^{-3}$ $m^3$ = litre. Thus an exponent affixed to a unit symbol containing a prefix indicates that the multiple or sub-multiple of the unit is raised to the power expressed by the exponent.

### Derived units

Derived units are related to the seven base units by definition. As we have seen, one important consequence of the SI is the establishing of a coherent unit of force, namely *the newton* (N), defined as that force which when applied to a body having a mass of one kilogram (kg) gives it an acceleration of one metre per second squared ($m/s^2$), i.e. N = kg $m/s^2$.

Derived units can themselves be used to form other derived units in a less cumbersome way than in terms of base units. For example the joule (J = N m = kg $m^2/s^2$) for energy, the watt (W = J/s) for power, and a deviation, namely the poise (P = $10^{-1}$ N $s/m^2$ or $10^{-1}$ Pa s) for dynamic viscosity, are given in the list on page 18. *However, it is essential to be able to translate special names of units into base units of the SI.*

### Special names and deviations from the SI

Although the CGPM recommends the use of SI units and their decimal multiples and sub-multiples, the CGPM and ISO have agreed (where it is thought, for convenience or practical considerations, that there is no compelling reason for changing) on special names and symbols for some of the derived units and on retaining certain well-established units (e.g. min, kW h) which deviate from the coherence of the SI. The deviations can, of course, be expressed in terms of SI units, if necessary; and the few imperfections and anomalies are a result of existing widespread usage, i.e. of a practical rather than a purely academic approach. For example, to have replaced the minute and the hour by decimal multiples of the second would have been 'purist' but hardly practical.

Hence it has been decided to retain:

(a) the usual larger units of time, i.e. minute (min = 60 s), hour (h = 3600 s), day and year.
(b) 360 degrees ($2\pi$ rad) in a circle because of international practice.
(c) the kilowatt hour (kW h = 3.6 MJ), which is a deviation from the SI because the hour is not a decimal multiple of the second.
(d) the litre as a special name for cubic decimetre (l = $dm^3$ = 1000 $cm^3$

$= 10^{-3}$ m$^3$) because it is a commonly used word for a unit volume in the commercial and domestic sectors (1 ml = 1 cm$^3$).

(e) the 'bar' for the multiple $10^5$ N/m$^2$ because (being equal to 750.1 mm Hg) it is nearly the previous 'standard atmosphere' (760 mm Hg) used as a unit of pressure in physics and chemistry, and the millibar (mbar = $10^2$ N/m$^2$ = 100 Pa) is used in meteorology throughout the world. For example, the atmospheric pressure which holds up the column of mercury of height $h$ in a barometer tube is given by

$$p_a = \rho_m gh = \frac{\rho_m}{\rho_w} (\rho_w g) h = d(\rho_w g) h$$

Thus, if $h$ = 760 mm,

$$p_a = 13.6 \times \left(\frac{10^3 \text{kg}}{\text{m}^3} \times \frac{9.807 \text{ m}}{\text{s}^2}\right) \times 0.76 \text{ m} \left[\frac{\text{N s}^2}{\text{kg m}}\right]$$

$$= 101.3 \text{ kN/m}^2 = 1.013 \text{ bar}$$

and 1 mbar = 0.075 mm Hg of barometer height.

(f) the name joule (J) for the unit of energy, defined as the work done when the point of application of a force of one newton is displaced through a distance of one metre in the direction of that force, i.e. J = N m. We may note here, however, that although a moment or torque has the same dimensions as energy (e.g. N m) this does not imply the same kind of physical quantity, as is the case vice versa (see page 63).

(g) the name hertz (Hz) for unit of frequency, defined as 'the frequency of a periodic phenomenon of which the periodic time is one second', i.e. the number of repetitions of a regular occurrence in one second. Engineers interpret the word 'frequency' according to the phenomena involved (e.g. rotational, angular and cyclic, whose customary modes of measurement are rev/s, rad/s and cycle/s respectively). Thus one kilocycle per second is nowadays said to be one kilohertz (1 kHz); and we may use 1 Hz as meaning 1 cycle/s = $2\pi$ rad/s in the case of a rotating crank or vector completing 1 cycle/revolution, in which case the unity multiplier is

$$\left[\frac{\text{s Hz}}{2\pi}\right] \quad \text{since [rad] is unity}$$

See Examples 14 and 15. Hz may, however, be used otherwise (e.g. the phenomenon of periodic blows of a drop-forge hammer may require Hz to be defined and used as meaning 1 blow/s, in which particular case [Hz s] = 1). Hence, to avoid errors due to ambiguous uses of hertz (if Hz is used at all) it must be given precise definition appropriate to the particular type of periodic phenomenon involved, since 'frequency' has no unique definition as a physical quantity (see Appendix 4). Because of this, just as the statement 'mass = 5 kg' is incomplete unless the question '5 kg of what?' can be answered, so 'frequency = 50 Hz or 50/s' requires an understanding (implied if not actually stated) of what recurs 50 times a second, i.e. of how precisely the unit, hertz, is being used. It may be used differently in different types of problem or specification.

As an example of the use of hertz, the statement that the frequency of revolution of the crankshaft of a four-stroke reciprocating-piston engine is 50 Hz implies that the shaft has a *rotational* frequency of 50 rev/s, i.e.

$$\left[\frac{\text{Hz s}}{\text{rev}}\right] = 1 = \left[\frac{\text{Hz s}}{2\pi\ \text{rad}}\right]$$

The *angular* frequency or angular velocity of the shaft is $100\pi$ rad/s which, if written $100\pi$ Hz, needs the qualifying note that in the case of *angular* frequency Hz is used as meaning rad/s, i.e.

$$\left[\frac{\text{Hz s}}{\text{rad}}\right] = 1$$

The *cyclic* frequency of the engine turning the crankshaft is 25 cycle/s, which could be written 25 Hz provided that it is realised that Hz is here used as meaning cycle/s, i.e.

$$\left[\frac{\text{Hz s}}{\text{cycle}}\right] = 1 = \left[\frac{\text{Hz s}}{2\ \text{rev}}\right] = \left[\frac{\text{Hz s}}{4\pi\ \text{rad}}\right]$$

in the case of a four-stroke engine, whereas for a two-stroke engine

$$\left[\frac{\text{Hz s}}{\text{cycle}}\right] = 1 = \left[\frac{\text{Hz s}}{1\ \text{rev}}\right] = \left[\frac{\text{Hz s}}{2\pi\ \text{rad}}\right]$$

Thus we have the possibility of hertz (Hz) meaning different things in one and the same mechanism, namely rev/s, rad/s, cycle/s, all of which are 'periodic phenomena of which the periodic time is one second'. *Hence the need for care and precise definition or specification of how the unit symbol, Hz, is being used (if at all) in each particular case.*

We may note that the familiar equation $\omega = 2\pi n = 2\pi f$ is a *numerical* equation or identity implying that one cycle is completed per complete turn or revolution of a crank or vector (see Appendix 4 and Examples 14 and 15).

*Special names and symbols for derived units and deviations*

| *Quantity* | *Letter symbol* | *Special name* | *Unit symbol* | *Expression in terms of SI units* |
|---|---|---|---|---|
| Time | $t$ | minute | min | 60 s |
| Volume | $V$ | litre | l | $dm^3 = 10^{-3} m^3$ |
| Mass | $m$ | tonne | t | $Mg = 10^3$ kg |
| Force | $F$ | newton | N | kg $m/s^2$ |
| Pressure | $P, p$ | pascal | Pa | $N/m^2$ = kg/m $s^2$ |
| | | bar | bar | $10^5$ $N/m^2$ |
| | | | 1 mm Hg | 133.3 Pa |
| Energy, work, quantity of heat | $E, W, Q$ | joule | J | N m = kg $m^2/s^2$ |
| | | kilowatt-hour | kW h | 3.6 MJ = 3.6 × $10^6$ kg $m^2/s^2$ |
| Power | $P$ | watt | W | J/s = kg $m^2/s^3$ |
| Dynamic viscosity | $\eta$ | poise | P | $10^{-1}$ kg/s m = dPa s |
| | | pascal-second | Pa s | kg/s m = 10 P = N $s/m^2$ |
| Kinematic viscosity | $\nu = \eta/\rho$ | stokes | St | $10^{-4}$ $m^2/s$ |
| Frequency | $f, \omega, n$ | hertz | Hz | 1/s (depending on what recurs in 1 s: cycle, rev, rad, impact, etc.) |
| Quantity of electricity | $Q$ | coulomb | C | A s |
| Electric resistance | $R$ | ohm | Ω | V/A = kg $m^2/(s^3$ $A^2)$ |
| Electric potential difference | $V = IR$ | volt | V | W/A = kg $m^2/(s^3$ A) |
| Electric capacitance | $C$ | farad | F | A s/V = $A^2$ $s^4$/(kg $m^2$) |
| Inductance | $L$ | henry | H | V s/A = kg $m^2/(A^2$ $s^2)$ |
| Magnetic flux | $\phi$ | weber | Wb | V s = kg $m^2/(A$ $s^2)$ |
| Luminous flux | $\phi$ | lumen | lm | cd sr } sr is not a base unit. It is a dimensionless ratio. |
| Illumination | $L$ | lux | lx | cd sr/$m^2$ } |

**Note** J can only replace N m when *energy* is concerned, as distinct from *moment or torque*—energy being a scalar product and torque a vector product of force and distance (see page 63).

The adjective 'specific' before the name of a physical quantity is restricted to meaning 'divided by mass' ($m$); thus the specific weight ($w/m$) of a mass $m$ is represented by $g$ (which changes with location), the specific weight of any mass on the earth being taken as 9.8067 N/kg (or 9.8067 $m/s^2$). Similarly, the word 'molar' means 'divided by amount of substance' (moles, $n$), e.g. an amount of substance $n$ having a volume $V$ and mass $m$ has a specific volume $v = V/m$, a molar mass $m/n = M$ and a molar volume $V_m = V/n$, where the subscript $m$ attached to the symbol for the extensive quantity $V$ denotes the corresponding molar quantity (see Example 22). Also, relative density $d = \rho/\rho_{(H_2O)} = \rho/\rho_w$ (see Example 6).

### Other derived units

As indicated in the following list, there are many derived units used without special names. Indeed some of the latter (e.g. bar and hertz) could probably well be abandoned in favour of base units of the SI which they represent and which it is essential to know, especially when reduction of clusters of units by cancellation of like with like is required.

| *Quantity* | *Letter symbol* | *Unit symbol (SI)* |
|---|---|---|
| Moment of inertia | $I$ | $kg\ m^2$ or $N\ m\ s^2$ |
| Second moment of area | $I$ | $m^4$ |
| Specific volume | $v = V/m$ | $m^3/kg$ |
| Density | $\rho = m/V$ | $kg/m^3$ |
| Velocity, linear | $u$ | m/s |
| Velocity, angular | $\omega$ | rad/s or 1/s |
| Torque | $T$ | N m (vector product) |
| Acceleration, linear | $a$ | $m/s^2$ |
| Acceleration, angular | $\alpha$ | $rad/s^2$ |
| Molar volume | $V_m = V/n$ | $m^3/mol$ |
| Molar mass | $M = m/n$ | kg/mol |
| Enthalpy, total | $H$ | $J = kg\ m^2/s^2$ |
| Enthalpy, specific | $h = H/m$ | $J/kg = m^2/s^2$ |
| Enthalpy, molar | $H_m = H/n$ | J/mol |
| Heat-capacity, total | $C$ | J/K |
| Heat-capacity, specific | $c = C/m$ | $J/kg\ K = m^2/s^2\ K$ |
| Heat-capacity, molar | $C_m = C/n$ | J/mol K |
| Thermal conductivity | $k$ | W/m K = J/s m K |
| Coefficient of heat-transfer | $h$ | $W/m^2\ K = kg/s^3\ K$ |
| Entropy, total | $S$ | J/K |
| Entropy, specific | $s = S/m$ | $J/kg\ K = m^2/s^2\ K$ |
| Entropy, molar | $S_m = S/n$ | J/mol K |
| Radiance | | $W/sr\ m^2$ |

Thermodynamic results from chemical or physical processes are expressed in equations for the process together with specification of the physical state—(g) gaseous, (l) liquid, (s) solid—of the participating substances and followed by the change in the appropriate thermodynamic function (enthalpy if the process is at constant pressure). For example,

$$H_2(g) + \tfrac{1}{2}O_2(g) = H_2O(l); \quad \Delta h_{298\,K} = -285.83 \text{ kJ/mol}$$

$$H_2O(l) = H_2O(g); \quad \Delta h_{298\,K} = +44.01 \text{ kJ/mol}$$

Since the mass of 1 mole of $H_2O$ is 18 g, the increase in specific enthalpy of $H_2O$ in changing phase from liquid (l) to dry saturated steam (g) at constant pressure and 298 K is

$$\Delta h_{298\,K} = 44.01 \frac{\text{kJ}}{\text{mol}} \left[\frac{\text{mol} \times 10^3}{18 \text{ kg}}\right]_{H_2O} = 2444 \text{ kJ/kg}$$

which is, of course, $h_{fg}$ in steam tables at 298 K or 24.85°C, the saturation temperature at 0.031 bar or 3.1 kPa, i.e. at 3100 $N/m^2$.

**Example 5**

Estimate the energy expected to be released when one milligram of fissile material is converted into energy.

Using Einstein's equation $\delta E = c^2 \delta m$, where $c$ represents the speed of light and $\delta m$ is one milligram (1 mg), we have

$$\delta E = (299.8 \text{ Mm/s})^2 \times (1 \text{ mg})$$

$$= (299.8)^2 \times 10^{12} \frac{\text{m}^2}{\text{s}^2} \times 10^{-3} \text{ g} \left[\frac{\text{kg}}{10^3 \text{ g}}\right] \left[\frac{\text{N s}^2}{\text{kg m}}\right]$$

$$= (299.8)^2 \times 10^6 \text{ N m} = 89.89 \times 10^9 \text{ N m or } 89.89 \text{ GJ}$$

**Example 6**

Neglecting losses, estimate the power required to pump $\dot{Q} = 3\ m^3/min$ of oil of relative density $d = 0.8$ from one tank to another against a difference of levels $H = 20$ m. Take the density of water as $10^3\ kg/m^3$ or 1 $Mg/m^3 = \rho_w$, hence density of oil is $\rho = d\rho_w = 0.8 \times 10^3\ kg/m^3$.

Neglecting friction and losses at bends, etc., the rate of working or the power required to transfer a fluid of density $\rho$ at a volumetric rate $\dot{Q}$ through a height $H$ is given by the physical formula

$$P = (\rho g)\,\dot{Q}H = 0.8 \times 10^3 \frac{\text{kg}}{\text{m}^3} \times 9.807\frac{\text{m}}{\text{s}^2} \times \frac{3\ \text{m}^3}{60\ \text{s}} \times 20\ \text{m} \left[\frac{\text{N s}^2}{\text{kg m}}\right]$$

$$= 7.846 \times 10^3\ \text{N m/s} = 7.846\ \text{kW}$$

**Example 7**

Calculate the force $F$ necessary to pull a smooth surface of area $A$ = 100 cm$^2$ at a speed of $u$ = 1 m/s relative to another if there is an oil film $y = \frac{1}{40}$ mm thick between the surfaces. The dynamic viscosity of the oil is $\eta$ = 20 P.

In general

$$\frac{F}{A} = \eta \frac{\mathrm{d}u}{\mathrm{d}y}$$

and, in this case, $F = A\eta u/y$. Hence

$$F = 100\ \text{cm}^2 \times 20 \left(10^{-1}\frac{\text{kg}}{\text{s m}}\right) \times \frac{1\ \text{m}}{\text{s}} \times \frac{40}{1\ \text{mm}} \left[\frac{10\ \text{mm}}{\text{cm}}\right]\left[\frac{\text{m}}{100\ \text{cm}}\right]$$

$$= 20 \times 40 \frac{\text{kg m}}{\text{s}^2} = 800\ \text{N}$$

**Example 8**

A mooring buoy weighs $W_B$ = 100 N and has a volume $V_B$ = 75 dm$^3$ or 75 litres. It is anchored by a vertical chain which weighs $W_c/L$ = 30 N/m. Estimate $L$, the greatest depth of anchor so that the buoy may not be pulled under the sea. The densities of sea water and the material of the chain are $\rho_s$ = 1.03 Mg/m$^3$ and $\rho_c$ = 7.7 Mg/m$^3$, respectively.

The force, $F$, exerted by the buoy on the top of the chain is equal and opposite to the force exerted by the chain on the buoy. Hence, if the buoy is just about to submerge,

$$F = (\text{buoyancy} - \text{weight})\ \text{ob buoy} = (\text{weight} - \text{buoyancy})\ \text{of chain}$$

$$= (B_B - W_B) = (W_c - B_c)$$

By Archimedes' principle

$$\text{Buoyancy} = \text{weight of water displaced}$$
$$= \text{mass of water displaced} \times g$$

Hence
$$B_B = \rho_s V_B g$$

and
$$B_c = \rho_s V_c g = \rho_s \left(\frac{W_c}{\rho_c g}\right) g = \frac{\rho_s}{\rho_c}\left(\frac{W_c}{L}\right) L$$

Thus
$$F = (\rho_s V_B g - W_B) = \left(\frac{W_c}{L}\right) L \left(1 - \frac{\rho_s}{\rho_c}\right)$$

or
$$1.03 \frac{\text{Mg}}{\text{m}^3} \times 0.075\ \text{m}^3 \times 9.807 \frac{\text{m}}{\text{s}^2}\left[\frac{\text{N s}^2}{\text{kg m}}\right] - 100\ \text{N}$$
$$= 30\ \text{N/m}\left(1 - \frac{1.03}{7.7}\right) \times L$$

i.e.
$$758\ \text{N} - 100\ \text{N} = \left(26 \frac{\text{N}}{\text{m}}\right) L \quad \text{and} \quad L = 25.3\ \text{m}$$

# 3

# Manipulation and reduction of units

## Dimensions and homogeneity

Certain physical quantities become intuitively familiar by the effects they produce on other quantities, e.g. force and acceleration, and temperature (first appreciated through our bodily senses). The word 'dimension' is used to indicate the *nature or kind of existence* of a quantity, irrespective of any system of units used to measure the size or amount of it. Thus, in this sense, 'dimension' is essentially intuitive, and refers to the kind or quality of a quantity under consideration. In particular, the words *mass, length* and *time* refer to things which we sense to be fundamentally different from each other. They may, therefore, be regarded as basic dimensions represented by (roman) capital letters [M], [L] and [T], these three being the minimum number of dimensions possible in mechanics. A volume $V$, for instance, can be conceived as a length cubed (or derived from length by double integration) and, hence, may be expressed dimensionally as $[\mathrm{L}^3]$, the square brackets being used to signify that only dimension, i.e. quality as distinct from quantity, is being considered. *The equation merely states that the dimension of a volume is equivalent to that of a length cubed.* Similarly, $[\mathrm{L/T}^2]$ does not specify an acceleration completely but merely its 'dimensions' in the sense of indicating the kind of units by means of which the magnitude of a linear acceleration can be expressed.

The 'dimensions' of many other physical quantities can be expressed in terms of the three fundamental dimensions: e.g. a force $F$ has dimensions $[\mathrm{ML/T}^2]$ as deduced from Newton's second law of motion, density $[\mathrm{M/L}^3]$, dynamic viscosity [M/LT], etc. The different forms of energy are of the same dimension $[\mathrm{ML}^2/\mathrm{T}^2]$ and may, therefore, be measured by means of one standard unit of energy, namely the joule. In the science of thermodynamics, temperature is recognised as an intensive property or function of state and as having a separate dimensional nature represented by [Θ], thus making four primary dimensions [M], [L], [T] and [Θ].

It is possible only for quantities of the same dimensions to be equated, added or subtracted. This is known as the condition of *dimensional homogeneity* and is the basis of analysis by the 'method of dimensions' which is much used in problems containing several variables (see Example 9: the requirement of unitary homogeneity). By dimensional analysis, dimensionless parameters or non-dimensional (Π) groups of variables can be composed for use as coordinates of graphs of experimental work (see page 25 and Example 21). Buckingham's pi (Π) theorem (which has nothing to do with $\pi = 3.1416\ldots$) states that if $p$ is the number of fundamental or base units required in expressing $n$ physical quantities, the latter can be made to yield $(n-p)$ dimensionless groups or Π terms such that the function $f(\Pi_1, \Pi_2, \ldots \Pi_{n-p}) = 0$.

Since physics relates quantities as entities, the equations expressing the relations must either equate quantities of the same dimensions or else establish relations between dimensionless quantities, i.e. numbers. Thus dimensional homogeneity implies that any physical equation can be expressed non-dimensionally since, if all the terms in it have the same dimensions, we have merely to divide through by any one of them to obtain a non-dimensional equation. A simple case is that of the well-known kinematic equation

$$s = ut + \tfrac{1}{2}at^2$$

which can be re-written as

$$\left(\frac{s}{ut}\right) = 1 + \tfrac{1}{2}\left(\frac{at}{u}\right)$$

The non-dimensional groups within the brackets may each be regarded as single variables or coefficients; and the latter equation can alternatively be written in the 'pi form' as

$$\Pi_1 = 1 + \tfrac{1}{2}\Pi_2 \quad \text{or} \quad \Pi_2 = f(\Pi_1)$$

where the function in this case is $f(\Pi_1) = 2\,(\Pi_1 - 1)$. This enables a graph to be drawn (*Figure 3.1(a)*) which displays all the information contained in the original equation. The graph can be used to find the acceleration for any given $s$, $u$ and $t$. Similarly,

$$\text{if} \quad \Pi_3 = (\Pi_1\Pi_2) = \left(\frac{sa}{u^2}\right) \quad \text{then} \quad \Pi_3 = \phi\,(\Pi_1)$$

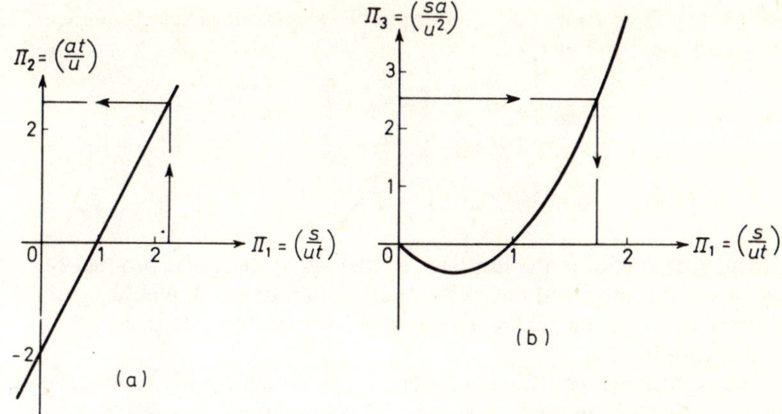


*Figure 3.1 Linear motion with constant acceleration*

where the function in this case is $\phi\,(\Pi_1) = 2\Pi_1\,(\Pi_1 - 1)$. This enables the graph (*Figure 3.1(b)*) to be drawn, which can be used to find the time taken for any given acceleration over any given distance and initial velocity.

The main importance of working in terms of non-dimensional parameters is, however, that it enables Π groups to be formed without a complete analysis of problems having many variables which defy theoretical solution. With the aid of graphs, solutions can be obtained from the measurements made in a limited number of experiments or tests in which there is no necessity to keep all variables constant (except the two to be plotted as abscissa and ordinate on graphs) as was the case prior to the method of dimensional analysis (also see Example 21). For example,

$$\left(\frac{gH}{\omega^2 D^2}\right) \text{ and } \left(\frac{\dot{Q}}{\omega D^3}\right)$$

are the non-dimensional parameters or coefficients which can be used as coordinates of graphs of observations on tests of axial-flow and centrifugal pumps whose blades rotate with angular velocity $\omega$.

**Example 9**

Although $E = 400$ J + 1 W h + 2 kN m is dimensionally correct, the numbers cannot be added until the measures or units are the same in each term. Thus, using 'unity multipliers' in square brackets for conversion purposes, we have

$$E = 400\ \mathrm{J} + 1\ \mathrm{W\,h}\left[\frac{\mathrm{J}}{\mathrm{W\,s}}\right]\left[\frac{3600\ \mathrm{s}}{\mathrm{h}}\right] + 2\ \mathrm{kN\,m}\left[\frac{\mathrm{J}}{\mathrm{N\,m}}\right]$$

$$= (400 + 3600 + 2000)\ \mathrm{J} = 6\ \mathrm{kJ}$$

provided that N m in the last term is the unit of the scalar product of force and distance, and not of the vector product which would signify a moment or torque, which itself is a vector quantity (see page 63 and Example 12).

Thus, although physical equations may be dimensionally homogeneous the numerics of the terms cannot be added or subtracted unless there is *unitary homogeneity* as well, i.e. until each number is associated with the same unit symbol.

**Conversion by unity multipliers**

It will be noticed that the quantities in the numerator and denominator of each of the square brackets (in all the Examples involving conversions of units) are equivalents. Hence, their ratios are unity [1] and, *since multiplying or dividing by unity does not alter the amount of anything,* appropriate unities or 'unity brackets' can be inserted either way up (as we have already seen) according to need and used as 'unity multipliers' any time a reduction or conversion to consistency or homogeneity of units is required. Although, in many cases, unit symbols can be crossed out and simply replaced by their equivalents, errors are certainly avoided if the appropriate unities are written down in square brackets before like units are cancelled with each other. To have written them down also enables subsequent checks to be made more easily on the quantitative and arithmetical work of problems, i.e. thought having been recorded on paper.

From this it follows that *the use of letter symbols in analysis to represent so-called 'conversion factors' is unnecessary and, indeed, wrong,* e.g. $J$ for the 'mechanical equivalent of heat or Joule's equivalent' in such energy equations as $E = JQ + W$ is unnecessary since $J$, used in this sense, is the ratio of two equal quantities of energy and is, therefore, *unity* as represented by, say,

$$\left[\frac{\mathrm{N\,m}}{\mathrm{J}}\right] \text{ or } \left[\frac{\mathrm{W\,s}}{\mathrm{J}}\right]$$

Thus, $J$ (not J) is not a physical quantity as are $Q$, $W$ and $E$, and it is misleading and wrong to insert it in physical equations since it is merely a particular 'unity multiplier' and only required when the last stage of a problem has been reached in which numbers and consistency of units (i.e. unitary homogeneity) are required (as e.g. in Example 9).

Similarly, it should be understood that a presupposition or assumption (often implied rather than explicitly stated) is that angles and functions of angles appearing in mathematical equations and customary formulae must be expressed in radians. Thus in all the usual formulae involving plane angle ($\theta$) and its ($\dot{\theta} = \omega$, $\ddot{\theta} = \alpha$) derivatives (e.g. $s = r\theta$; $W = T\theta$; $P = T\omega$; $E = \frac{1}{2} I\omega^2$; $T = I\alpha$) plane angles are presupposed expressed in radians; i.e. $\theta$, $\omega$, $\alpha$ are 'radian-restricted' because of the advantages of circular measure in analysis generally (see Appendix 2). It follows that the usual letter symbols representing plane angle and its derivatives express *numbers* of radians. This is equivalent to writing (in accordance with the Stroud convention) numbers × rad (rad/s or rad/s$^2$, as the case may be ), where rad = 1 (unity), *'radian' being a mere name of the angle whose circular measure is unity.* However, although the circular measure of a plane angle, being the ratio of like quantities, is dimensionless (i.e. is a pure number) it is necessary in quantitative work to insert rad as an indicator (answering the question 'number of what?') of how plane angles have been quantified. Hence if, as is customary, angles and angular motion are first expressed by means of degrees or revolutions, conversion to radian measure can be achieved by using the 'unity multipliers'

$$\left[\frac{\pi \text{ rad}}{180 \text{ deg}}\right] = 1 = \left[\frac{2\pi \text{ rad}}{\text{rev}}\right]$$

to be followed by replacing rad by unity at any convenient point (i.e., in effect, omitting rad) in the final arithmetical stage of a problem involving formulae deduced on the assumption or presupposition that plane angles are expressed in radians.

Other 'unity brackets' resulting either from definitions or equivalent measurements of like physical quantities are:

$$\left[\frac{60 \text{ s}}{\text{min}}\right], \left[\frac{\text{N s}^2}{\text{kg m}}\right], \left[\frac{\text{N m}}{\text{J}}\right], \left[\frac{\text{W s}}{\text{N m}}\right], \left[\frac{\text{J s}^2}{\text{kg m}^2}\right], \left[\frac{\text{knot s}}{0.514 \text{ m}}\right],$$

$$\left[\frac{3.6 \text{ MJ}}{\text{kW h}}\right], \left[\frac{\text{N}}{\text{Pa m}^2}\right], \left[\frac{1 \text{mm Hg}}{133.3 \text{ Pa}}\right], \left[\frac{10 \text{ bar}}{1 \text{ MPa}}\right], \left[\frac{760 \text{ mm Hg}}{\text{std atmos}}\right],$$

$$\left[\frac{\text{std atmos m}^2}{101.325 \text{ kN}}\right], \left[\frac{\text{m}^3}{10^3 \text{ litre}}\right], \left[\frac{\text{ml}}{\text{cm}^3}\right], \left[\frac{\text{mol}}{0.012 \text{ kg}}\right]_{\text{C}},$$

$$\left[\frac{\text{mol}}{28.84 \text{ g}}\right]_{\text{air}}, \left[\frac{\text{Pa s}}{10 \text{ P}}\right], \left[\frac{10 \text{ P s m}}{\text{kg}}\right], \left[\frac{\text{rad}}{1}\right] \text{ etc.}$$

**Example 10**

A ship of mass 10 000 tonnes or 10 million kilograms increases its speed uniformly from 10 to 21.7 knots in 10 min, calculate the resultant force (thrust of propellers – resistance of water) required to achieve this.

The uniform acceleration $a = \dfrac{11.7 \text{ knot}}{10 \text{ min}} = \dfrac{1}{51.4}$ knot/s

Hence the resultant force required is

$$F = ma = 10^7 \text{ kg} \times \frac{1}{51.4} \frac{\text{knot}}{\text{s}} \left[\frac{0.514 \text{ m}}{\text{knot s}}\right] \left[\frac{\text{N s}^2}{\text{kg m}}\right]$$

$$= 10^5 \text{ N or } 0.1 \text{ MN}$$

**Example 11**

Calculate the kinetic energy of a gyroscope wheel of moment of inertia $I = 1.2$ Mg mm$^2$ (or 1.2 mN m s$^2$) spinning with a speed $\omega =$ 20 000 rev/min.

$$\text{Kinetic energy} = \tfrac{1}{2} I\omega^2 = \frac{1.2}{2} \text{ Mg mm}^2 \left(2 \times 10^4 \frac{\text{rev}}{\text{min}}\right)^2$$

$$= 2.4 \times 10^{11} \text{kg} \frac{\text{m}^2}{10^6} \frac{\text{rev}^2}{3600 \text{ s}^2} \left[\frac{4\pi^2}{\text{rev}^2}\right] \left[\frac{\text{N s}^2}{\text{kg m}}\right]$$

$$= \frac{8 \times 10^8}{3 \times 10^6} \pi^2 \text{ N m} = 2.632 \text{ kJ.}$$

This example shows that it is not necessary to use different symbols ($\omega$ and $n$) for angular velocity and rotational speed, *provided* that units are written down and reduced by appropriate unity multipliers, including [rad] = 1, which therefore, as here, can be omitted.

**Example 12**

A gyroscope wheel of mass $\frac{3}{4}$ kg and radius of gyration 20 mm is found to process at the rate of 72 degree/h when rotating at a speed of 18 000 rev/min. Estimate the couple causing the precession, using the physical formula $C = I\Omega\omega$, where $C$ represents the moment or couple causing an angular rate of precession $\omega$ to the axis of a wheel rotating at an angular rate $\Omega$ and of moment of inertia $I$ about the axis of spin.

Using 'unity multipliers' in square brackets for the purpose of reduction of units, we confidently calculate

$$C = (\tfrac{3}{4} \times 400)\,\text{kg}\left(\frac{\text{m}}{10^3}\right)^2 \times \frac{18\,000\text{ rev}}{60\text{ s}} \times \frac{1.2\text{ deg}}{60\text{ s}}\left[\frac{\pi}{180\text{ deg}}\right]$$

$$\times\left[\frac{2\pi}{\text{rev}}\right]\left[\frac{\text{N s}^2}{\text{kg m}}\right] = \frac{\pi^2\text{ N m}}{50\,000} = 0.197\text{ mN m} \quad\text{or}\quad 0.197\text{ N mm}$$

which is not mJ because $C$ is a torque, not energy (see Example 9).

**Example 13**

Air enters a jet engine at the rate of 20 kg/s in an aeroplane flying at 432 knots or 800 km/h. The mass air/fuel ratio is 60 and the velocity of the gas *relative* to the tail-duct is 2400 km/h. Calculate the thrust and power developed in flight, and the overall thermal efficiency if the energy released by (or 'calorific value' of) the fuel is 46 MJ/kg.

Thrust = rate of change of momentum. Hence, referring to *Figure 3.2*, we have

$$F = (\dot{m}_a + \dot{m}_f)\,u_2 - \dot{m}_a u_1 = \dot{m}_a(u_2 - u_1) + \dot{m}_f u_2$$

$$= 20\frac{\text{kg}}{\text{s}} \times \frac{1600\text{ km}}{3600\text{ s}} + \frac{20\text{ kg}}{60\text{ s}} \times \frac{2400\text{ km}}{3600\text{ s}}$$

$$= \frac{20}{36}(16 + 0.4)\frac{\text{kg km}}{\text{s}^2}\left[\frac{\text{N s}^2}{\text{kg m}}\right]$$

$$= 9.12\text{ kN}$$

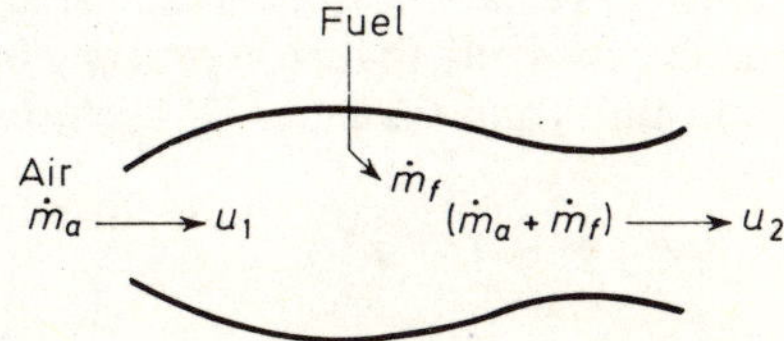


*Figure 3.2 Fluid flow through a jet engine*

$$\text{Power} = \text{thrust} \times \text{speed of aeroplane}$$

$$= 9.12 \text{ kN} \times \frac{800 \text{ km}}{3600 \text{ s}} \left[\frac{\text{kW s}}{\text{kN m}}\right]$$

$$= 2.03 \text{ MW}$$

$$\text{Potential power in fuel input} = \frac{20 \text{ kg}}{60 \text{ s}} \times 46 \frac{\text{MJ}}{\text{kg}} \left[\frac{\text{MW s}}{\text{MJ}}\right]$$

$$= 15.3 \text{ MW}$$

$$\therefore \text{Overall thermal efficiency} = \frac{2.03}{15.3} = 0.133 \text{ or } 13.3\%$$

**Simple harmonic motion**

Rectilinear oscillatory motion in which the acceleration is negatively proportional to the displacement is called *simple harmonic motion.* Mathematically this can be written

$$\ddot{x} \propto (-x) \quad \text{or} \quad \ddot{x} = -\omega^2 x$$

where $\omega$ is a constant, as yet not identified. Alternatively,

$$\frac{d^2x}{dt^2} + \omega^2 x = \left[D^2 + \omega^2\right] x = 0$$

which equation is satisfied by

$$x = A \sin(\omega t + \beta)$$

where $A$ and $\beta$ are arbitrary constants, necessarily, required in the solution of a second-order differential equation. $A$ is the amplitude in linear measure; $\beta$ is a plane angle in radian measure and depends on the initial ($t = 0$) conditions; $\omega$ is the constant angular frequency in circular measure (i.e. radians per unit time) and $t$ is time. This solution is verified by

$$\dot{x} = \omega A \cos(\omega t + \beta)$$

and

$$\ddot{x} = -\omega^2 A \sin(\omega t + \beta) = -\omega^2 x$$

Thus it follows that s.h.m., being a sine (or cosine) function of time, can also be represented as the length of the diametral projection ($x$) of the tip of a vector of length $A$ rotating on a circular path with constant angular velocity or circular frequency $\omega$ necessarily in radians per unit time (see Example 14(b)) because of the presupposition implicit in the mathematics.

An alternative solution of $[D^2 + \omega^2]\ x = 0$ is in the exponential form

$$x = A_1 \epsilon^{j\omega t} + A_2 \epsilon^{-j\omega t}$$

where $A_1$ and $A_2$ are constants and $j = \sqrt{-1}$. Since $\epsilon^{j\omega t} = \cos \omega t + j \sin \omega t$ and $\epsilon^{-j\omega t} = \cos \omega t - t \sin \omega t$, this solution transforms, of course, into

$$x = A_3 \sin \omega t + A_4 \cos \omega t = A \sin (\omega t + \beta)$$

the previous trigonometrical solution.

**Example 14**

(a) What is the use of idealised systems of mechanical vibrations?

(b) Analyse the simplest of these, namely that of a mass $m$ oscillating freely (i.e. undamped natural oscillation) in rectilinear motion on the end of a helical spring of stiffness $k$. Show that this s.h.m. can be represented by a sine or cosine function of time.

(c) An instrument of mass $m = 10$ kg is set on three rubber mounts, each having a deflection of 0.75 mm/N. Calculate the cyclic frequency, the periodic time of the system when oscillating, and the circular frequency of a corresponding rotating vector. Neglect the effect of the mass of the spring.

(a) Complicated vibrating mechanical systems are often analysed in the first instance by replacing them with simpler equivalent systems of block masses, springs and damping devices in order to be able to subject them to mathematical analysis; and although the solutions of the differential equations of motion of such idealised systems only approximate to reality, measurements on experimental and analogous systems (electrical as well as mechanical) suggest modifications which enable the near-performance of an actual system to be predicted. The actual performance can, of course, only be revealed with the aid of recording and measuring instruments placed in appropriate places on the real system.

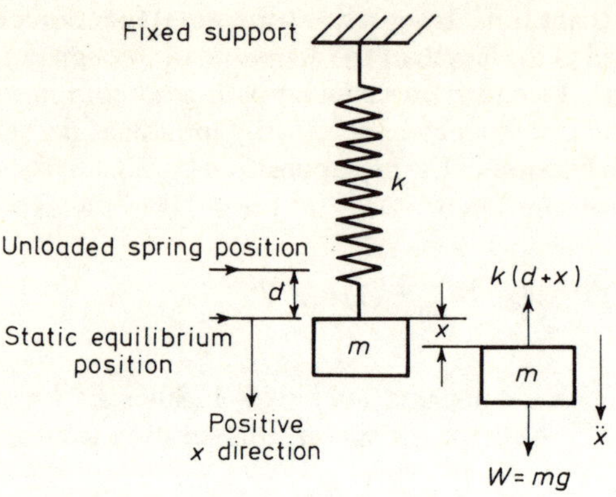


*Figure 3.3 Simple harmonic motion*

(b) Referring to *Figure 3.3*, if the static extension of the loaded spring is $d$, then the equation of motion (in accordance with Newton's second law, $F = ma$) when the mass is displaced a distance $x$ from its static-equilibrium position is

$$mg - k(d + x) = m\ddot{x}$$

or, since $mg = kd$,

$$\frac{d^2x}{dt^2} = -\left(\frac{k}{m}\right)x \quad \text{or} \quad [D^2 + (k/m)]\; x = 0$$

By trial,

$$x = A \sin \omega t + B \cos \omega t = X \sin(\omega t + \phi)$$

proves to be a solution to the differential equation because $d^2x/dt^2 = -\omega^2 x$ and $X = \sqrt{(A^2 + B^2)}$, and $\phi = \tan^{-1}(B/A)$. Hence, if $x = X \sin(\omega t + \phi)$ is to represent the s.h.m., it follows that

$$\omega = \sqrt{(k/m)} \quad \text{which has dimensions } [T^{-1}]$$

In emphasis of the fact that such mathematical solutions presuppose plane ($\theta$) and its derivative ($\theta = \omega$) to be in circular measure (i.e. the units of $\omega$ are, necessarily, rad/unit of time), this may be written more clearly as (see also Appendix 4):

$$\frac{\dot{\theta}}{\text{rad}} = \frac{\omega}{\text{rad}} = \sqrt{(k/m)} \quad \text{or} \quad \omega = \sqrt{(k/m)}\ \text{rad}$$

The cycle of motion is completed at intervals of $2\pi$ rad $= \omega\tau$; hence the periodic time is

$$\tau = \frac{2\pi}{\omega}\ \text{rad} = \frac{1\ \text{cycle}}{f}$$

where we should be clear that $\tau$, $\omega$ and $f$ represent *not* mere numbers, but *either* physical quantities (see Chapter 4) whose magnitudes are expressed by numbers × units, *or* modes of measurement (usually s, rad/s and cycle/s, respectively; see Appendix 4). Hence the cyclic frequency may be written

$$f = \frac{\text{cycle}}{\tau} = \frac{1}{2\pi}\sqrt{(k/m)}\ \text{cycle} = \frac{1}{2\pi}\sqrt{(g/d)}\ \text{cycle}$$

If the starting or initial conditions of this s.h.m. as represented by $x = X \sin(\omega t + \phi)$ are $t = 0$, $x = X$ and $\dot{x} = 0$, then $X = X \sin\phi$ and $x = 0 = \omega X \cos\phi$; hence

$$\phi = \frac{\pi}{2}\ \text{rad} \quad \text{and} \quad x = X \sin\left(\omega t + \frac{\pi}{2}\right) = X \cos \omega t$$

(c) The stiffness of the three rubber mounts taken together is

$$k = \frac{3\text{N}}{0.75\ \text{mm}} = 4\ \text{N/mm}$$

Hence, the angular frequency of the system is

$$\omega = \sqrt{(k/m)}\ \text{rad} = \sqrt{\left(4 \times 10^3 \frac{\text{N}}{\text{m}} \times \frac{1}{10\ \text{kg}}\left[\frac{\text{kg m}}{\text{N s}^2}\right]\right)}\ \text{rad}$$

$$= 20 \frac{\text{rad}}{\text{s}}$$

which, since in this case

$$\left[\frac{\text{cycle}}{2\pi\ \text{rad}}\right] = 1$$

converts into a cyclic frequency

$$f = 20 \frac{\text{rad}}{\text{s}} \left[\frac{\text{cycle}}{2\pi\ \text{rad}}\right] = 3.18 \text{ cycle/s or } 3.18 \text{ Hz}$$

where Hz is used as cycle/s (or $2\pi$ rad/s of a corresponding rotating vector). The periodic time is

$$\tau = \frac{\text{cycle}}{f} = \frac{\pi}{10} \text{ s} = 0.314 \text{ s}$$

one cycle being completed at intervals of 0.314 s.

We may note that it is really unnecessary to use two letter symbols, $\omega$ and $f$, *so long as the ways of measurement,* cycle and radian ( a dimensionless quantity) being customary modes of measurement rather than units in the strict sense, *are written alongside the corresponding numerical values.* For example, periodic time $\tau$ is the inverse of cyclic frequency, i.e.

$$\tau = \frac{\text{cycle}}{f} = \frac{\text{cycle}}{3.18 \text{ Hz}} \left[\frac{\text{Hz s}}{\text{cycle}}\right] = 0.314 \text{ s}$$

This could alternatively be obtained from the *numerical* formula

$$|\tau| = \frac{1}{|f|} = \frac{2\pi}{|\omega|}$$

(see Appendix 4), where $\tau = |\tau|$ s, $f = |f|$ cycle/s, and $\omega = |\omega|$ rad/s. Thus

$$|\tau| = \frac{1}{3.18} = \frac{2\pi}{20} = 0.314$$

i.e.

$$\tau = 0.314 \text{ s}$$

For purposes of conversion and reduction of units, therefore, it is advisable to follow the procedure of inserting units (and customary ways of measurement such as rad, rev, cycle) and appropriate ‘unity multipliers’ based on unique definitions, e.g.

$$\left[\frac{2\pi\ \text{rad}}{\text{rev}}\right] = 1$$

and, to suit each problem, particular definitions or usages, e.g.

$$\left[\frac{\text{Hz s}}{\text{cycle}}\right] = 1 = \left[\frac{\text{cycle}}{2\pi \text{ rad}}\right]$$

in the case of Hz and cycle. This procedure always safeguards against errors in quantitative work, and also provides for a relatively easy double-check on such work.

**Example 15**

A mass $m$ suspended on a light helical spring of stiffness $k$ vibrates with a natural cyclic frequency of 1.6 Hz. With an extra 0.5 kg added to the mass, the frequency is reduced to 1.3 Hz, where Hz represents cycle/s or $2\pi$ rad/s of a rotating vector. Calculate the mass $m$ and the stiffness of the spring. State any assumptions made.

We may use this particular example to show that the solution of engineering problems in general divides into four sections, namely:

*Assumptions:* (a) the effect of the mass of the spring is negligible; (b) the spring operates within its elastic limit; (c) air resistance and inherent damping is negligible; (d) the spring is suspended from a rigid support; (e) the motion of the mass is constrained to a single degree of freedom, rectilinear in this case.

*Scientific laws:* (a) Hooke's law applies to the spring; (b) Newton's second law of motion applies to the mass.

*Mathematics:*

$$F = ma \quad \text{or} \quad -kx = m\frac{\mathrm{d}^2x}{\mathrm{d}t^2}$$

where $x$ is the displacement of the mass (*Figure 3.3*) from the static-equilibrium position, and $k$ is the spring stiffness. Hence,

$$\frac{\mathrm{d}^2x}{\mathrm{d}t^2} + \omega^2 x = 0$$

where

$$\left(\frac{\omega}{\text{rad}}\right) = \sqrt{(k/m)}$$

*Arithmetic:* in the first case, the frequency

$$\omega_1 = \sqrt{(k/m)}\ \text{rad} = 10.05 \text{ rad/s or } 1.6 \text{ Hz}$$

and in the second case,

$$\omega_2 = \sqrt{\left(\frac{k}{m + 0.5\ \mathrm{kg}}\right)}\ \mathrm{rad} = 8.17\ \mathrm{rad/s}\ \text{or}\ 1.3\ \mathrm{Hz}$$

Eliminating $k$, we get

$$\frac{m + 0.5\ \mathrm{kg}}{m} = \left(\frac{\omega_1}{\omega_2}\right)^2 = \left(\frac{1.6}{1.3}\right)^2 = 1.513$$

Hence

$$m = \frac{0.5\ \mathrm{kg}}{0.513} = 0.975\ \mathrm{kg}$$

and, converting Hz by

$$\left[\frac{2\pi\ \mathrm{rad}}{\mathrm{Hz\ s}}\right] \text{ and rad } = 1$$

we deduce that

$$k = m\left(\frac{\omega_1}{\mathrm{rad}}\right)^2 = 0.975\ \mathrm{kg}\left\{1.6\ \mathrm{Hz}\left[\frac{2\pi}{\mathrm{Hz\ s}}\right]\right\}^2\left[\frac{\mathrm{N\ s^2}}{\mathrm{kg\ m}}\right]$$

$$= 98.5\ \mathrm{N/m}\ \text{or}\ 0.985\ \mathrm{N/cm}$$

These results of theory can, of course, easily be verified by observations on simple apparatus in a laboratory.

**Further applications**

The following worked examples show the manipulation of units in the mechanics of fluids, dimensional analysis, combustion and heat transfer.

**Example 16**

Calculate the dynamic pressure caused by a wind of 100 km/h striking the surface of a large building normally so as to lose all its momentum. Assume the static pressure of the air to be 0.1 MPa (or 1 bar) and the temperature 16.85°C.

Force $F = \dot{m}\Delta u$, where $m = \rho a u$ and $\Delta u = u$. Hence the dynamic pressure exerted is

$$p = \frac{F}{a} = \rho u^2$$

Density of air is

$$\rho = \frac{m}{V} = \frac{P}{RT} = \frac{1\ \text{bar}\left[\dfrac{10^5\ \text{N}}{\text{bar m}^2}\right]}{287\ \dfrac{\text{J}}{\text{kg K}} \times 290\ \text{K}\left[\dfrac{\text{N m}}{\text{J}}\right]} = \frac{10}{8.323}\ \text{kg/m}^3$$

Hence dynamic pressure is

$$\begin{aligned} p &= \rho u^2 \\ &= \left(\frac{10}{8.323}\ \frac{\text{kg}}{\text{m}^3}\right)\left(\frac{100\ \text{m}}{3.6\ \text{s}}\right)^2\left[\frac{\text{N s}^2}{\text{kg m}}\right] \\ &= \frac{10^5}{107.9}\ \text{N/m}^2 = 9.27\ \text{m bar or } 927\ \text{Pa} \end{aligned}$$

**Example 17**

(a) Derive Bernoulli's equation for the steady flow of an incompressible, inviscid fluid. Use it to estimate the speed of air flowing at low speed through an open-ended horizontal wind-tunnel at a section where a water-manometer records $h$ = 25 mm of vacuum, the density of the air being $\rho_a$ = 1.288 kg/m$^3$ and that of water $\rho_w$ = 1 tonne/m$^3$.

(b) Estimate the dynamic pressure on the nose or at a 'stagnation point' on the wing of an aeroplane travelling at 720 knots at a height where the density of the stationary air is $\rho_a$ = 1.266 kg/m$^3$

The four sections (assumptions, scientific laws, mathematics, arithmetic), generally involved in the solution of engineering problems, will be clearly seen.

(a) *Assumptions:* the fluid is (a) incompressible (i.e. density $\rho_a$ is constant), (b) inviscid (i.e. dynamic viscosity $\eta = 0$ or viscous drag is zero), (c) in steady flow (i.e. the conditions at each point i· the fluid do not change with time, or $(\partial u/\partial t)_s = 0$ at each cross-section), (d) flowing adiabatically (i.e. there is no heat-transfer, $\delta Q = 0$).
*Scientific laws:* Newton's second law of motion applies to an element of this ideal fluid because as each element passes fron one point to the next it must, in general, accelerate so as to ar with the velocity appropriate to that point.

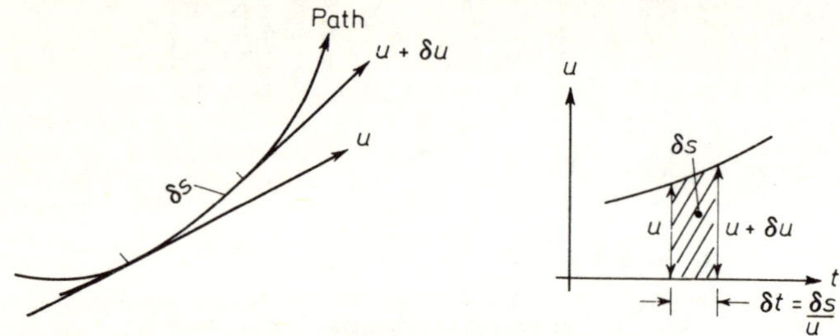


*Figure 3.4 Acceleration of a particle*

*Mathematics:* referring to *Figure 3.4*, each element changes its speed by an amount $\delta u$ whilst travelling the distance $\delta s$, which is represented by the shaded area under the speed-time curve (see *Figure 5.1*). Hence, if $\delta s$ is sufficiently small, we may write $\delta s = u\delta t$. The corresponding acceleration of the element (*Figure 3.4*) along the path in the $s$ direction is

$$a_s = \frac{\delta u}{\delta t} = u\frac{\delta u}{\delta s}$$

since, in steady flow, $u$ is only a function of $s$ and not of $t$ as well. We may note that, in general, $u = f(s, t)$ and hence, in the notation of partial differentials,

$$\frac{du}{dt} = \frac{\partial u}{\partial s}\frac{ds}{dt} + \frac{\partial u}{\partial t}\frac{dt}{dt}$$

but, for steady flow, $\partial u/\partial t = 0$ and hence

$$\frac{du}{dt} = \frac{ds}{dt}\left(\frac{\partial u}{\partial s}\right)_t = u\frac{du}{ds}$$

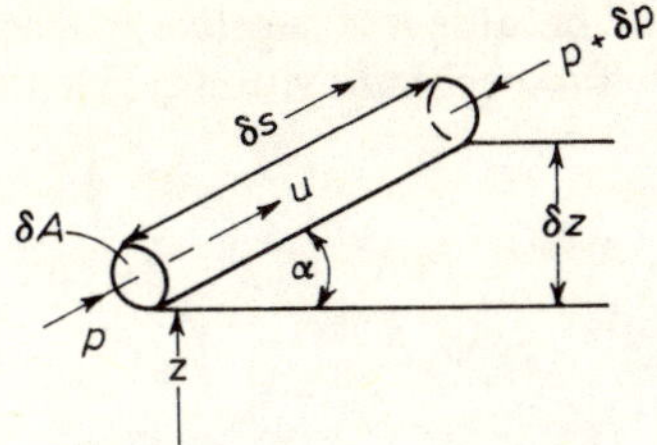


*Figure 3.5 An element of fluid*

Thus referring to *Figure 3.5* we see that, in the direction of motion, the force on the mass $\rho\delta A\delta s$ is $-\delta A\delta p - (\rho\delta A\delta s)\, g \sin\alpha$, i.e. $-(\delta p + \rho g\,\delta z)\,\delta A$. Hence

$$\delta p + \rho g\,\delta z + \rho u\delta u = 0$$

or $$\frac{\delta p}{\rho} + \delta\left(\frac{u^2}{2}\right) + g\delta z = 0 \quad \text{or} \quad \frac{p}{\rho} + \frac{u^2}{2} + gz = \text{constant}$$

for any particular $g$. This is known as Bernoulli's equation.
We may note that, instead of applying Newton's second law in the form $F = ma$, we could have used the form $F = \dot{m}\,\delta u$ to get

$$-(\delta p + \rho g\delta z)\,\delta A = (\rho u\delta A)\,\delta u,$$

i.e. $$\delta p + \rho g\delta z + \rho u\delta u = 0$$

as before.
*Arithmetic:* Assuming that the air is drawn from rest in the laboratory space into the wind-tunnel, the velocity of the air at any cross-section of the tunnel is given by

$$\tfrac{1}{2}u^2 = (p_{atm} - p)/\rho_a$$

where the pressure at the section is given by

$$p = p_{atm} + \Delta p \quad \text{and} \quad \Delta p = \rho_w g h$$

(i.e. below atmospheric because $h$ is negative). Hence the velocity of air passing through the specified section is given by

$$u^2/2 = -\Delta p/\rho_a$$

i.e.

$$u = \sqrt{(-2gh\rho_w/\rho_a)}$$

$$= \sqrt{\left\{2 \times 9.807\frac{\text{m}}{\text{s}^2}\left(\frac{25}{10^3}\text{m}\right)\left(\frac{10^3}{1.288}\right)\right\}}$$

$$= 19.5 \text{ m/s or } 38 \text{ knots}$$

(b) The dynamic pressure $(p - p_{atm})$ deduced from equation

$$\frac{p_{atm}}{\rho} + \frac{u^2}{2} = \frac{p}{\rho}$$

(considering the air oncoming relative to the aircraft) is

$$p - p_{atm} = \tfrac{1}{2}\rho u^2$$

$$= \tfrac{1}{2} \times 1.226\frac{\text{kg}}{\text{m}^3} \times (720 \text{ knot})^2 \left[\frac{(0.514)^2\text{m}^2}{\text{knot}^2\ \text{s}^2}\right]\left[\frac{\text{N s}^2}{\text{kg m}}\right]$$

$$= 84 \text{ kN/m}^2 \text{ or } 84 \text{ kPa}$$

**Example 18**

Estimate the exit diameter of a convergent nozzle, shaped such that its coefficient of discharge is unity, to deliver water at the rate $\dot{m}$ = 48 kg/min when the drop in pressure over the nozzle is $\Delta p$ = 1 MN/m$^2$. Neglect the velocity of approach and take the density of water as 1 Mg/m$^3$ or 1 t/m$^3$.

The velocity $u$ of liquid leaving the nozzle is given by

$$\frac{u^2}{2} = \frac{\Delta p}{\rho}$$

(where $\Delta p$ is the pressure decrement) as deduced from Bernoulli's equation or from the steady-flow energy equation. Hence

$$\dot{m} = \rho A u = A\sqrt{(2\rho\Delta p)}$$

and $$A = \frac{48\ \text{kg}}{60\ \text{s}} \frac{1}{\sqrt{\left\{2000\ \frac{\text{kg}}{\text{m}^3} \times 10^6\ \frac{\text{N}}{\text{m}^2} \left[\frac{\text{kg m}}{\text{N s}^2}\right]\right\}}}$$

$$= \frac{4\ \text{kg}}{5\ \text{s}} \left(\frac{1}{\sqrt{(2 \times 10^9)}} \times \frac{\text{m}^2\ \text{s}}{\text{kg}}\right) = \frac{4\ \text{m}^2}{22.35 \times 10^4}$$

and exit diameter is

$$d = \frac{40\ \text{mm}}{\sqrt{(\pi \times 22.35)}} = 4.78\ \text{mm}$$

**Example 19**

Calculate the velocity of sound in air at 16.85°C, assuming $\gamma = 1.4$ and using (a) the characteristic or specific gas constant for air is $R$ = 287 J/kg K, (b) the universal molar gas constant is $\bar{R}$ = 8.3143 J/mol K. Velocity of sound in gas is given by the physical formula $a = \sqrt{(\gamma RT)}$.

(a) The Celsius temperature 16.85°C implies a thermodynamic temperature $T$ = (273.15 + 16.85) K = 290 K. Hence

$$a^2 = \gamma RT = 1.4 \times 287\ \frac{\text{J}}{\text{kg K}} \times 290\ \text{K} \left[\frac{\text{N m}}{\text{J}}\right] \left[\frac{\text{kg m}}{\text{N s}^2}\right]$$

$$= 116\,520\ \text{m}^2/\text{s}^2\text{, i.e. } a = 341\ \text{m/s}$$

(b) Assuming air to contain 21% oxygen and 79% nitrogen by volume, and since a volumetric analysis ($P$, $T$, $R$ constant, i.e. $n = (P/RT)V$) is also a mole analysis (see Appendix 5):
1 mol of air = 0.21 $O_2$ + 0.79 $N_2$ has a mass (0.21 × 32 + 0.79 × 28) g = 28.84 g. Hence

$$a^2 = \gamma RT$$

$$= 1.4 \times 8.3143\ \frac{\text{N m}}{\text{mol K}} \times 290\ \text{K} \left[\frac{\text{mol}}{0.02884\ \text{kg}}\right] \left[\frac{\text{kg m}}{\text{N s}^2}\right]$$

$$= 116\,520\ \text{m}^2/\text{s}^2\text{, i.e. } a = 341\ \text{m/s}$$

This shows that insertion and reduction of units prevents an error in using the wrong $R$ (see page 43, and also page 62).

**Example 20**

Air flows along a duct to enter a convergent–divergent nozzle with a temperature 833 K and a velocity 200 m/s. Reversible adiabatic (isentropic) expansion would reduce its temperature to 433 K. Calculate the velocity and Mach number which would result at the exit if reversible adiabatic expansion could take place along the nozzle according to the law $PV^\gamma$ = constant. For air $\gamma = 1.4$, specific heat-capacity at constant pressure is $c_p = 1.005$ kJ/kg K and the specific gas constant for air is $R = 0.287$ kJ/kg K.

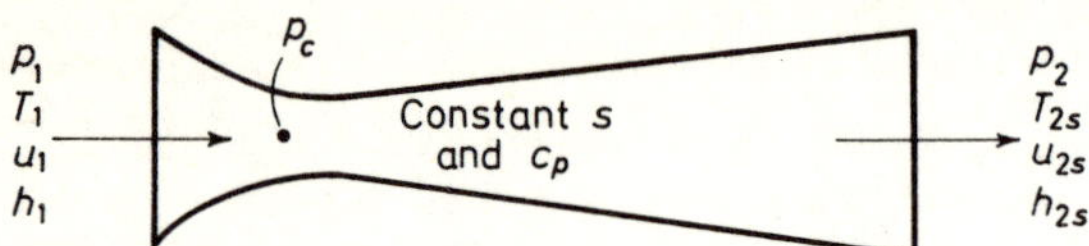


*Figure 3.6 Convergent–divergent nozzle*

If expansion could be reversible (i.e. frictionless and without eddies) as well as being adiabatic (i.e. heat-insulated) the physical equation (see also Example 33) for this particular (hypothetical) case of *reversible* adiabatic (constant entropy *s*, i.e. isentropic) flow may, referring to *Figure 3.6*, be written:

$$u_{2s}^2 - u_1^2 = 2(h_1 - h_{2s}) = 2c_p\,(T_1 - T_{2s})$$

where the suffix (2*s*) denotes the state after reversible adiabatic (constant *s*) expansion, *u* represents velocity, *T* temperature and *h* specific enthalpy.

Hence

$$u_{2s}^2 - \left(200\frac{\text{m}}{\text{s}}\right)^2 = 2 \times 1.005\ \text{kJ/kg K} \times (833 - 433)\ \text{K}$$

i.e. $$u_{2s}^2 = 4\frac{10^4\ \text{m}^2}{\text{s}^2} + 2.01 \times 400\frac{\text{kN m}}{\text{kg}}\left[\frac{\text{kg km}}{\text{kN s}^2}\right]$$

$$= 4\,\frac{10^4\ \text{m}^2}{\text{s}^2} + 80.4\,\frac{10^4\ \text{m}^2}{\text{s}^2} = 84.4\,\frac{10^4\ \text{m}^2}{\text{s}^2}$$

Thus the exit velocity would be $u_{2s} = 920$ m/s. The velocity of sound in air at exit temperature $T_{2s}$ would be

$$a_{2s} = \sqrt{(\gamma R T_{2s})} = \sqrt{\left\{1.4 \times 287 \frac{\text{J}}{\text{kg K}} \times 433 \text{ K} \left[\frac{\text{kg m}}{\text{N s}^2}\right] \left[\frac{\text{N m}}{\text{J}}\right]\right\}}$$

$$= \sqrt{\left(1.4 \times 287 \times 433 \frac{\text{m}^2}{\text{s}^2}\right)} = 417 \text{ m/s}$$

Hence the Mach number (i.e. the ratio of the air velocity to the local velocity of sound) at exit, after reversible adiabatic (isentropic) expansion, would be $Ma = \frac{920}{417} = 2.2$.

We may note in passing that there is a limit to the rate at which gas can flow through a nozzle, as determined by the condition for 'choked flow':

$$p_c/p_1 = \left(\frac{2}{\gamma + 1}\right)^{\gamma/\gamma - 1}$$

where $p_c$ is the (critical) pressure at the throat (for air, $p_c = 0.528\, p_1$) which determines the maximum flow rate however low $p_2$ may be. Also, the velocity through the throat is

$$v_c = \sqrt{(\gamma R T_c)}$$
$$= \sqrt{(\gamma p_c/\rho_c)}$$

which is at Mach 1 there.

**Example 21**

Estimate the minimum diameter $D$ of particles (assumed spherical) of emery powder which, after being scattered on the surface of water 480 mm high in a glass jar, are found to have cleared the water and settled on the bottom in 100 s. The relative density of emery is $d_e$ = 4.0 and the dynamic viscosity of the water is $\eta = 0.012$ P and density $\rho_w = 10^3$ kg/m$^3$.

Applying the method of dimensions or dimensional analysis (see page 23) we may assume that the drag force $F$ on a body moving through an incompressible fluid depends on the density ($\rho$) and the dynamic viscosity ($\eta$) of the fluid, and on the size ($l$) and speed ($u$) of the relative motion between the body and the fluid; i.e. the variable force $F$ depends on, or is a function of, the four independent variables, each of which could be changed irrespective of the others. Thus we start by assuming $F = \phi(\rho, \eta, l, u)$. A typical term in a power series will be $C\rho^a\eta^b l^c u^d$, in which $C, a, b, c$ and $d$ are, as yet, unspecified numbers. Dimensional homogeneity requires that

$$\left[\frac{\mathrm{ML}}{\mathrm{T}^2}\right] = \left[\frac{\mathrm{M}}{\mathrm{L}^3}\right]^a \left[\frac{\mathrm{M}}{\mathrm{LT}}\right]^b \left[\mathrm{L}\right]^c \left[\frac{\mathrm{L}}{\mathrm{T}}\right]^d$$

Hence equating the indices of M, L and T results in the three equations $1 = a + b$, $1 = -3a - b + c + d$ and $2 = b + d$, which involve four unknowns and are therefore incapable of complete solution. Three of the unknowns can, however, be expressed in terms of the fourth, e.g. $a = 1 - b$, $d = 2 - b$, $c = 1 + 3a - d + b = 2 - b$. Hence the expression for $F$ involves terms of the type:

$$C\left(\frac{\rho}{\rho^b}\right)\left(\eta^b\right)\left(\frac{l^2}{l^b}\right)\left(\frac{u^2}{u^b}\right) \quad \text{or} \quad C\rho l^2 u^2 \left(\frac{\rho l u}{\eta}\right)^{-b}$$

in which $C$ and $b$ may take any numerical values without violating the requirement of dimensional homogeneity. Thus the right hand side is simply $C\rho l^2 u^2$ times some unspecified function ($f$) of $(\rho l u/\eta)$, i.e.

$$\frac{F}{\rho l^2 u^2} = f\left(\frac{\rho l u}{\eta}\right)$$

which can, alternatively, be written

$$\frac{F}{\frac{1}{2}\rho u^2 A} = \phi\left(\frac{\rho u l}{\eta}\right) = C_D$$

or as coefficient of drag $C_D = \phi(Re)$, where $F$ is the resultant force on the body of plan area $A$ and linear size $l$ (diameter $D$ in the case of a sphere), $\rho$ and $\eta$ are the density and dynamic viscosity of the fluid, respectively, $u$ is the velocity of the body relative to the fluid and $\frac{1}{2}\rho u^2$ is the dynamic pressure of the stream (see Example 17).

The right-hand side of the equation denotes some unspecified function ($\phi$) of Reynolds number ($Re$). Thus, although dimensional analysis is able to predict that $C_D$ is uniquely related to $Re$, it is unable to state the exact form of the relationship between them. This relationship is usually revealed by means of a graph after plotting $C_D$ against $Re$ for a series of tests in which any of the variables may take on different values. The tests yield points which fall on one graph, and thus the effect of a number of variables can be judged from one unique curve. Hence the importance of dimensional analysis and value of non-dimensional plotting. The omission of a relevant variable in the original assumption will cause scatter of points on the graph, and hence necessitate a re-analysis with the missing variable included.

In the case of a sphere falling with constant velocity through a fluid (the 'terminal' velocity being when the drag force of the liquid is equal to the force due to gravity less buoyancy of the sphere) the graph of $C_D$ against $Re$ is a hyperbola such that $C_D \times Re = 24$, or

$$F = \tfrac{1}{2}\rho u^2 \times \frac{\pi}{4} D^2 \times 24 \left(\frac{\eta}{\rho u D}\right) = 3\pi\eta u D$$

(known as Stokes' law, since he deduced the latter term mathematically before Rayleigh devised the method of dimensions).

The resultant downward force on a spherical particle of emery of diameter $D$ is the difference between its weight and buoyancy, i.e.

$$F = W - B = \frac{\pi D^3}{6} g(\rho_e - \rho_w) = \frac{\pi D^3}{6} g\rho_w(d_e - 1)$$

Hence $F = 3\pi\eta u D = \dfrac{\pi D^3}{6} g\rho_w(d_e - 1)$

and $$D^2 = \frac{18\eta u}{\rho_w(d_e - 1)g} = \frac{18 \times 0.0012 \dfrac{\text{kg}}{\text{m s}} \times \dfrac{48 \text{ m}}{10^4 \text{ s}}}{10^3 \dfrac{\text{kg}}{\text{m}^3} \times 3 \times 9.81 \dfrac{\text{m}}{\text{s}^2}} = \frac{35}{10^{10}} \text{m}^2$$

Thus the minimum diameter of a spherical particle of this emery is

$$D = 5.92 \frac{\text{m}}{10^5} = 59.2 \ \mu\text{m} \text{ or } 0.0592 \text{ mm}$$

**Example 22**

Calculate the molar volume (i.e. the volume of one mole) of *any* gas at temperature $T = 273.15$ K (0°C) and pressure 101.325 kN/m$^2$ (760 mm Hg).

Molar volume of gas is $V_m = V/n = \bar{R}T/P$, i.e. one mole of any gas (obeying $P\rho = RT$) occupies the same volume at the same temperature and pressure. In this (s.t.p.) case

$$V_m = 8.3143 \frac{\text{N m}}{\text{mol K}} \times \frac{273.15 \text{ K m}^2}{101.325 \text{ kN}} = 2.2418 \times 10^{-2} \text{m}^3/\text{mol}$$

$$= 22.42 \text{ litre/mol}$$

Also it may be noted that, at the same temperature and pressure, the volume of gas is proportional to the 'amount of substance' (moles), since $V = (\bar{R}T/P)\,n$. (See also Example 34 and Appendix 5.)

**Example 23**

If the products of combustion of the liquid fuel $C_7H_{16}$ show equal volumes of $CO_2$ and $O_2$ with no CO, calculate

(a) the air/fuel ratio of the stoichiometric mixture,

(b) the actual air/fuel ratio,

(c) the percentage excess air, assuming air to contain 21% oxygen and 79% nitrogen by volume, i.e. 28.84 g/mol (see Example 19).

(a) Since the molar mass (see Appendix 5) of the fuel is $M = 7 \times 12 + 16 \times 1 = 100$ g/mol, i.e. 1 mol of this fuel has a mass 100 g, then the chemical equation for the combustion of 1 mol of fuel requiring $x$ mol of air for combustion of the stoichiometric mixture may be written:

$$C_7H_{16} + x\,(0.21\,O_2 + 0.79\,N_2) = a\,CO_2 + b\,H_2O + d\,N_2$$

Hence, on equating the *number of corresponding atoms,* we obtain

$$a = 7 \text{ mol of } CO_2;\; b = 8 \text{ mol of } H_2O;$$

$$a + \tfrac{1}{2}b = 0.21\,x = 11 \text{ mol}$$

i.e. $x = 52.4$ mol and $d = 0.79\,x = 41.4$ mol.

Hence the stoichiometric air/fuel ratio is $x = 52.4$ mol of air per mol of fuel, or

$$\frac{52.4 \text{ mol air}}{1 \text{ mol fuel}}\left[\frac{1 \text{ mol fuel}}{100 \text{ g fuel}}\right]\left[\frac{28.84 \text{ g air}}{1 \text{ mol air}}\right] = 15.1 \text{ kg air/kg fuel}$$

(b) If, in practice, $y$ mol of air are supplied per 1 mol (i.e. 100 g) of fuel, the combustion equation may be written

$$C_7H_{16} + y(0.21\,O_2 + 0.79\,N_2) = aCO_2 + bH_2O + dO_2 + eN_2$$

from which, on equating the number of corresponding atoms, we obtain

$$a = 7 \text{ mol};\; b = 8 \text{ mol};\; d + \tfrac{1}{2}b + a = 0.21\,y = d + 11$$

and, since the analysis of the products gave $d = a = 7$ mol, then $0.21\,y = 7 + 11 = 18$ mol. Hence, $y = 85.71$ mol of air per mol of fuel as the actual air/fuel ratio, or

$$y = 0.8571\frac{\text{mol air}}{\text{g fuel}}\left[\frac{28.84\text{ g air}}{1\text{ mol air}}\right] = 24.8\text{ kg air/kg fuel}$$

(c) Thus the excess air is 24.8 – 15.1 = 9.7 kg/kg of fuel, or 33.31 mol air/mol fuel, i.e. 64% in excess of the air required for the stoichiometric mixture.

**Example 24**

Although the masses of reactants and products of combustion remain the same, show that the process of combustion of a hydrocarbon changes not only the kind but also the number of molecules and moles. Also, show that the number of gaseous moles may be different from the number of molecules.

Combustion of the general hydrocarbon, $C_\alpha H_\beta$, in a minimum of air can be represented by the equation:

$$C_\alpha H_\beta + \left(\alpha + \frac{\beta}{4}\right)\left\{O_2 + \frac{79}{21}N_2\right\} = \alpha\, CO_2 + \frac{\beta}{2}H_2O + \frac{79}{21}\left(\alpha + \frac{\beta}{4}\right)N_2$$

from which we see that there is a molecular contraction (i.e. a reduction in the number of molecules) of $1 - \beta/4$. Also, if the hydrocarbon fuel is in the liquid phase and the $H_2O$ in the products is allowed or induced to condense, the number of gaseous moles reduces by $\beta/4$.

**Example 25**

Find (a) the percentage molecular change in a stoichiometric air/petrol mixture, assuming the petrol to be represented by the chemical symbol $C_8H_{18}$, (b) the minimum volume (at s.t.p.) and mass of air needed to burn 1 litre of petrol of relative density 0.75, i.e. of density $\rho = 0.75$ Mg/m$^3$ or $\frac{3}{4}$ t/m$^3$, and (c) the amount of water in the products of combustion.

(a) $1C_8H_{18} + (8 + 4.5)O_2 + 47\,N_2 = 8CO_2 + 9H_2O + 47\,N_2$

i.e. 60.5 molecules of reactants become 64 molecules of products, which corresponds to a molecular expansion of 3.5

molecules or 5.8 per cent, but the mass before and after combustion remains the same. Also

$$1C_8H_{18}\ (l) + 59.5\ \text{Air}\ (g) = 8\ CO_2\ (g) + 47\ N_2\ (g) + 9H_2O\ (l)$$

i.e. the number of gaseous moles reduces by 59.5 – 55 = 4.5 moles.

(b) One mole of $C_8H_{18}$ or (96 + 18) g = 0.114 kg needs a minimum of 12.5 mol of $O_2$ associated with 47 mol of $N_2$ in air, and 0.114 kg of $C_8H_{18}$ needs a minimum of 12.5 × 32 g or 0.40 kg of $O_2$ and 47 × 28 g or 1.316 kg $N_2$ for complete combustion; i.e. 1 mol of $C_8H_{18}$ petrol needs a minimum of 59.5 mol of air or, expressed in terms of mass, 0.114 kg of petrol needs 1.716 kg of air.

Hence 1 litre of $C_8H_{18}$ liquid petrol, having a mass ($\rho V$) of 0.75 Mg/m$^3$ × 10$^{-3}$ m$^3$ = 0.75 kg, requires a minimum of (0.75/0.114) × 1.716 kg = 11.28 kg of air, or (0.75/0.114) × 59.5 = 391 mol of air which corresponds (see Example 22) to a volume of air (at s.t.p.) of

$$V = nV_m = 391 \times 22.4\ \text{litre} = 8\ 800\ \text{litre or}\ 8.8\ \text{m}^3$$

(since it is customary to use 'litre' only when a substance is in the liquid phase).

(c) Also 1 mol of $C_8H_{18}$ liquid petrol yields 9 mol of $H_2O$, which condenses to water if the products of combustion are exhausted into the atmosphere. Hence 1 litre of this liquid petrol of mass 0.75 kg yields

$$9 \times \frac{18}{114} \times 0.75\ \text{kg} = 1.067\ \text{kg of water}$$

or, since $V = m/\rho$,

$$\frac{1.067\ \text{kg}}{1.0\ \text{kg/litre}} = 1.067\ \text{litre of water}$$

**Example 26**

Cooling water flows at the rate of 3 Mg/min through a bank of 100 tubes in a condenser. Each tube has a length 5 m, an external diameter 25 mm and a wall thickness 1.25 mm. Steam condenses on the outer surfaces of the tubes. Calculate the heat-transfer coefficient on the water side of the tubes, and find the pressure of the condensing steam if the heat-transfer coefficient on the steam side is 11.35 kW/m$^2$ K. The rise in temperature of the cooling water is 5 K and its mean temperature is 25°C.

From tables we find that, for water at 25°C, $\rho = 1.0\ \text{Mg/m}^3$, $c_p = 4.18\ \text{kJ/kg K}$, $\eta = 890 \times 10^{-6}\ \text{kg/m s}$ and $k = 611 \times 10^{-6}\ \text{kW/m K}$.

Hence the mean velocity of water in the tubes is $u = \dot{m}_1/\rho a$, where $\dot{m}_1$ is the rate of flow through one tube of cross sectional area $a$, i.e.,

$$u = \frac{10^{-3}}{2}\frac{\text{Mg}}{\text{s}} \times \frac{\text{m}^3}{1\ \text{Mg}} \times \frac{4}{\pi}\left(\frac{1000}{22.5\ \text{m}}\right)^2 = 1.257\ \text{m/s}$$

The Reynolds number, $Re$, is

$$\frac{uD}{\nu} = \frac{uD\rho}{\eta} = \left(1.257\frac{\text{m}}{\text{s}}\right)(22.5 \times 10^{-3}\text{m})\left(\frac{10^3\ \text{kg}}{\text{m}^3}\right)\left(\frac{\text{m s}}{890 \times 10^{-6}\text{kg}}\right)$$

$$= 31\,400$$

The Prandtl number, $Pr$, is

$$\frac{\eta c_p}{k} = \left(890 \times 10^{-6}\frac{\text{kg}}{\text{m s}}\right)\left(4.18\frac{\text{kJ}}{\text{kg K}}\right)\left(\frac{10^6\ \text{m K s}}{611\ \text{kJ}}\right)$$

$$= 6.10$$

The Nusselt number, $(Nu) = 0.023\ Re^{0.8}\ Pr^{0.4} = 187.7$, i.e.

$$(Nu) = \frac{hD}{k} = 187.7$$

Hence the coefficient of heat-transfer on the water side is

$$\frac{(Nu)k}{D} = h = 187.7 \times \left(611 \times 10^{-6}\frac{\text{kW}}{\text{m K}}\right)\left(\frac{10^3}{22.5\ \text{m}}\right)$$

$$= 5.1\ \text{kW/m}^2\ \text{K}$$

The overall heat-transfer coefficient, $U$, based on the surface area, $A$, of the steam side, is given by

$$\frac{1}{U} = \frac{1}{h_o} + \frac{1}{h_i}\left(\frac{r_o}{r_i}\right)$$

where $h_o$ and $h_i$ are the heat-transfer coefficients for the outer and inner surfaces with radii $r_o$ and $r_i$, i.e.

$$\frac{1}{U} = \left(\frac{1}{11.35} + \frac{1}{5.1 \times 0.9}\right) \frac{\text{m}^2\ \text{K}}{\text{kW}} \quad \text{or } U = 3.07\ \text{kW/m}^2\ \text{K}$$

and the rate of heat-transfer $q = AU\Delta T_m = \dot{m}c_p\Delta T_w$. Hence the difference in the mean temperatures of the steam and the water is

$$\Delta T_m = T_s - T_w$$

$$= \left(\frac{10^2}{2}\frac{\text{kg}}{\text{s}}\right)\left(4.18\frac{\text{kJ}}{\text{kg K}}\right)(5\ \text{K})\left(\frac{1000}{\pi \times 25 \times 500\ \text{m}^2}\right)\left(\frac{\text{m}^2\ \text{s K}}{3.07\ \text{kJ}}\right)$$

$$= 8.68\ \text{K}$$

and $t_s$ = 8.68 K + 25 Celsius = 33.68°C or 306.83 K, which is the saturation temperature at pressure 0.053 bar or 5.3 kN/m$^2$.

4

# Physical and numerical formulae

## Use of letter symbols in formulae

At the outset, in order to be quite clear on the difference between physical and numerical formulae and their use, it will be helpful first to recapitulate on the use of symbols.

We have seen in previous chapters how in analysis, letter symbols such as $m, g, t, v, \omega, \rho, P, V, T, \phi$ may be used to represent either *numbers* or *physical quantities* according to convention or assumptions. They appear in *numerical* or *physical formulae*, respectively, and the standard measures or units associated with each number in the former are specified at the beginning of the analysis.

The convention of using letter symbols to represent physical quantities (sometimes referred to as the Stroud convention) is quantity algebra: physical quantities cannot be measured or specified merely by numbers but only by comparison with like quantities. Therefore, since physics is concerned with inanimate nature and its laws, a physical quantity is specified by means of a *unit* or *measure* (of the same nature as the quantity being measured) and a *number* showing how many times that unit is contained in the given quantity (a *sign* such as → or – indicating direction or sense may need to be attached in the case of vector quantities such as force, torque and velocity: see page 63). The term *physical quantity* is used to denote any quantity composed of *a number* × *a unit* so as not to confuse with the single generic word 'quantity'.

Using the Stroud convention we have, for example,

(a) a length $L$ of 6 m, not a length of $L$ m,

(b) a time $t$ = 8 s, not $t$ s, for in the latter case $t$ would represent only the *number* 8,

(c) an acceleration $g$ of 9.81 $m/s^2$, not an acceleration of $g$ $m/s^2$.

Thus *a letter symbol used to denote a physical quantity does not imply the choice of a particular unit* (see Appendix 2 for the one exception: plane angle).

The quotient, however, of the symbol for a physical quantity and its unit-symbol is a pure number. For example, a thermodynamic temperature $T = 305$ K could be written $T/\mathrm{K} = 305$, and such ratios of letter symbols to unit symbols ($L$/m, $t$/s, etc.) could well be used as labels of the numbers on the axes of graphs plotted as a result of experiments concerning variations in physical quantities.

Thus quantity algebra in which letter symbols are used to represent physical quantities is freed of parallel thought concerning particular units; this only arises when the quantitative or arithmetical stage (see Examples 14 and 17) of a problem has been reached, and necessarily requires *each letter symbol to be replaced by its numerical value* × *the unit* used to determine the numeric. Such algebraic analyses result in **physical formulae** or equations which state a balance between certain physical quantities; e.g. force $F = ma$, torque $T = I\alpha$, kinetic energy $E = \frac{1}{2}mu^2$, velocity of sound in gas $a = \sqrt{(\gamma RT)} = \sqrt{(\gamma p/\rho)}$, velocity of incompressible fluid through nozzle $u = \sqrt{(2(-\Delta p/\rho))}$, etc., in which the signs, numbers and their units are contained in the letter symbols. We may represent the transference of energy by the process of heat-transfer by, say, $Q = -3$ J, meaning that 3 joules of energy have left a certain body or system. The sign can often be rendered unnecessary by using suffixes; e.g. $Q_i$ and $Q_o$ represent inflow and outflow, respectively, from a 'system' (say, an energy transformer such as an engine or a refrigerator etc.) within an imaginary (or systems) boundary (see *Figure 4.1*).

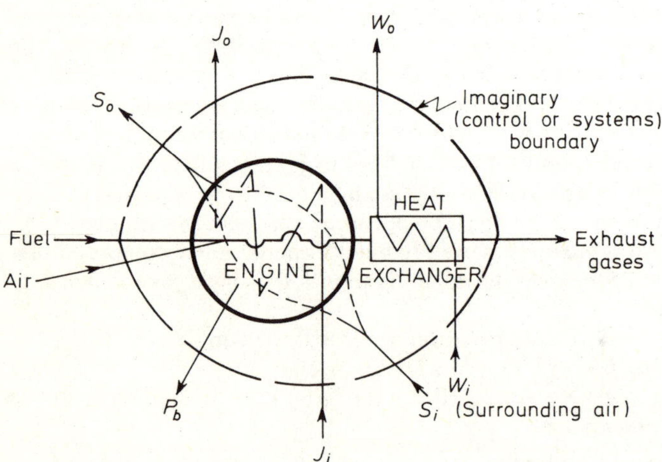


*Figure 4.1 Energy transformer*

In **numerical formulae** letter symbols represent *numbers* resulting from *measurements in particular units.* Examples are:

(a) empirical formulae, usually deduced from graphs as a result of experiments; e.g. $v = 0.159z^{5/2}$ for the flow of water over a 90° vee notch, $v$ being in litres/hour and $z$, the head, in millimetres (see Example 30).

(b) theoretically deduced numerical formulae; e.g. the velocity of gas flowing adiabatically through a nozzle from a large vessel (i.e. $u_1 = 0$) is $u_2 = 44.7\sqrt{(h_1 - h_2)}$, where $u_2$ is in m/s and the reduction in specific enthalpy $(h_1 - h_2)$ is in kJ/kg (see Example 33).

All numerical formulae are restricted or bound by particular units whereas physical formulae are universal in that they are free from the restriction or attachment of any particular units or system of units.

*Although units are often omitted (presumably with the idea of saving time) it is essential, if errors are to be avoided, especially when groups of units are involved, that students should not only be dimension-conscious but realise that the best check on units is made by always inserting them alongside the numerical values of physical quantities and then proceeding to cancel like units with each other – a practice especially desirable in the early stages of courses in engineering science.*

As in the case of any other convention, the Stroud convention has to be tried for its advantages to be appreciated.

**Example 27**

Convert the physical formula $P = T\omega$ for power transmitted by a rotating shaft into a numerical formula in which power $P$ is in kilowatts when torque $T$ is measured in newton metres and the shaft speed or rotational frequency is measured in revolutions per minute. What power is transmitted by a shaft rotated at 3000 rev/min by a torque of 280 N m?

The physical formula for power transmitted may be written $P = T\omega$, where $T$ represents torque (i.e. a number × a unit) and $\omega$ represents angular speed of the shaft – usually thought of in radians per unit of time but, as we shall now see, it could equally well be in revolutions per unit time. In fact a *single* letter symbol (say, $\omega$) representing the physical quantity rotational speed, angular velocity, rotational or angular frequency, can be used *provided that*, when quantitative work becomes involved, the units are explicitly written in, using rad = 1.

The particular units of kilowatt (kW) for power, newton metre (N m) for torque, and revolutions per minute (rev/min) can be imposed on this physical formula by rewriting it in the form

$$\left(\frac{P}{\text{kW}}\right) = \left(\frac{T}{\text{N m}}\right) \times \left(\frac{\omega}{\text{rev/min}}\right) \left\{\frac{\text{N m rev}}{\text{kW min}}\right\} \left[\frac{\text{J}}{\text{N m}}\right] \left[\frac{\text{W s}}{\text{J}}\right] \left[\frac{2\pi}{\text{rev}}\right] \left[\frac{\text{min}}{60\text{ s}}\right]$$

using appropriate unity brackets to reduce the formula to

$$b = tn \frac{2\pi}{60\,000} \quad \text{or } b = \frac{\pi tn}{30\,000}$$

where $b$, $t$ and $n$ are *numbers* ($n$ is the *number* of rev/min, $t$ is the *number* of N m of torque and $b$ is the *number* of kW transmitted), i.e. where $P = b$ kW; $T = t$ N m and $\omega = n$ rev/min. Alternatively,

$$\left(\frac{P}{\text{W}}\right) = \frac{\pi}{30} \left(\frac{T}{\text{N m}}\right) \left(\frac{\omega}{\text{rev/min}}\right) = \left(\frac{T}{\text{N m}}\right) \left(\frac{\omega}{\text{rad/s}}\right)$$

Thus, using the first *numerical* formula, namely $b = \pi tn/30\,000$, corresponding to the physical formula $P = T\omega$, if $t = 280$ and $n = 3000$, then

$$b = \frac{\pi \times 280 \times 3000}{30\,000} = 88 \quad \text{i.e. } P = 88 \text{ kW}$$

Alternatively, using the *physical* formula straight away, we get

$$P = T\omega = 280 \text{ N m} \times \frac{3000 \text{ rev}}{60 \text{ s}} \left[\frac{2\pi \text{ rad}}{\text{rev}}\right]$$

$$= 28\pi \times 10^3 \text{ N m/s} = 88 \text{ kW}$$

**Example 28**

Calculate the torque, the angle of twist per metre of length and the maximum shear stress on a hollow steel shaft of internal and external diameters $d = 100$ mm and $D = 300$ mm respectively, when transmitting a power of 3 MW at 140 rev/min. Assume a shear modulus $G = 80$ GN/m$^2$.

Using the physical formula of Example 27, namely $P = T\omega$, we deduce that

$$T = \frac{P}{\omega} = (3 \text{ MN m}) \left(\frac{60 \text{ s}}{140 \text{ rev}}\right) \left[\frac{\text{rev}}{2\pi}\right]$$

$$= \frac{9}{14\,\pi} \text{ MN m} = 0.205 \text{ MNm}$$

Since the second polar moment of area is

$$J = \frac{\pi}{32}(D^4 - d^4) = \frac{\pi}{32}(D^2 - d^2)(D^2 + d^2)$$

i.e. $$J = \frac{\pi}{32}(8 \times 10^9)\ \text{mm}^4 = \frac{\pi}{4} \times 10^{-3}\ \text{m}^4$$

then

$$\frac{\theta}{L} = \frac{T}{GJ} = \frac{0.205\ \text{MN m}}{80\ \text{GN/m}^2 \times \frac{\pi}{4} \times 10^{-3}\ \text{m}^4} = \frac{4}{\pi} \times \frac{0.205\ \text{GN m}}{80\ \text{GN m}^2}$$

$$= \frac{0.01025}{\pi\ \text{m}}\left[\frac{180\ \text{deg}}{\pi}\right] = 0.185\ \text{deg/m}$$

Also, since $$\frac{T}{J} = \frac{\tau}{r} = \frac{G\theta}{L}$$

the maximum shear stress is

$$\tau_{max} = \frac{T}{J}\frac{D}{2}$$

i.e. $$\tau_{max} = \frac{0.205\ \text{MN m}}{\frac{\pi}{4} \times 10^{-3}\ \text{m}^4} \times \frac{0.30}{2}\ \text{m}$$

$$= 39.2\ \text{MN/m}^2 \text{ or } 39.2\ \text{MPa}$$

**Example 29**

Calculate the viscous shear stress on a flat plate when air of viscosity $\eta = 17.9$ mg/s m or 179 $\mu$P flows parallel to the plate with a velocity distribution given by the numerical formula $u = 40(y - \frac{5}{3}y^2)$, in which $u$ is the number of m/s and $y$ is the distance from the plate in mm.

The corresponding physical formula giving the velocity distribution is

$$\frac{U}{\text{m/s}} = 40\left(\frac{Y}{\text{mm}} - \frac{1}{6}\frac{Y^2}{\text{mm}^2}\right) = \frac{40}{\text{mm}}\,Y\left(1 - \frac{1}{6\ \text{mm}}Y\right)$$

where the capital letter symbols represent physical quantities (numbers × units), i.e.

$$U = \frac{40\,000}{s} Y\left(1 - \frac{1}{6\text{ mm}} Y\right)$$

Hence $\quad \dfrac{dU}{dY} = \dfrac{40\,000}{s}\left(1 - \dfrac{1}{3\text{ mm}} Y\right) = \dfrac{40\,000}{s} \qquad$ when $Y = 0$

Therefore on the plate the shear stress is

$$\tau = \eta \frac{dU}{dY} = 17.9 \frac{\text{mg}}{\text{m} \times \text{s}} \; \frac{40\,000}{\text{s}} \left[\frac{\text{N} \times \text{s}^2}{10^6 \text{ mg} \times \text{m}}\right] = 0.716 \text{ N/m}^2$$

**Change of units of numerical formulae**

The method for changing the units of numerical (including empirical) formulae from one specified set to another is composed of three stages (see Examples 29 and 30):

(a) change the numerical formula in which letter symbols (say, lower case) represent numbers (i.e. numerics) only, into one in which the different letter symbols (say, capitals) represent physical quantities (i.e. numbers × units);

(b) change the units of this latter formula by operating on the unit symbols with appropriate unity multipliers or 'unity brackets';

(c) change this formula, in which the letter symbols (capitals) represent physical quantities, back into its corresponding numerical formula, in which letter symbols (lower case) represent numbers or numerics resulting from using the required units.

**Example 30**

As a result of observations and measurements made during a test of the flow of water over a 90-degree vee notch, a graph was plotted of the quantity $q$ in cubic metres per minute against the head $h$ over the notch in centimetres, from which graph the empirical law

$$1000\,q = 0.84\,h^{5/2}$$

was deduced. Convert this law into the corresponding empirical law which gives the rate of flow in litres per second when the head over the notch is measured in millimetres, and calculate the rate of flow when the head is 100 mm.

(a) So as not to confuse l as the unit symbol for litre with figure one, the name litre is written in full. Let $\dot{Q} = q\ \text{m}^3/\text{min} = v$ litre/s, where $\dot{Q}$ is a physical quantity (number × unit), and $q$ and $v$ are merely numbers. Also, let $H = h$ cm $= z$ mm, where $H$ represents a physical quantity, and $h$ and $z$ numerals.

The given empirical (i.e. numerical) formula

$$1000\,q = 0.84\,h^{5/2}$$

can be rewritten

$$\left(\frac{\dot{Q}}{\text{m}^3/\text{min}}\right) = \frac{0.84}{1000}\left(\frac{H}{\text{cm}}\right)^{5/2}$$

or

$$\dot{Q} = \left(\frac{0.84}{1000}\ \frac{\text{m}^3}{\text{min cm}^{5/2}}\right) H^{5/2}$$

in which it will be noticed that the so-called 'constant' within the brackets depends on the units used.

(b) Appropriate unity brackets enable the rewritten formula to be converted for use with the required units, as follows:

$$\left(\frac{\dot{Q}}{\frac{\text{m}^3}{\text{min}}\left[\frac{\text{min}}{60\ \text{s}}\right]\left[\frac{\text{litre}}{10^{-3}\ \text{m}^3}\right]}\right) = \frac{0.84}{1000}\left(\frac{H}{\text{cm}\left[\frac{10\ \text{mm}}{\text{cm}}\right]}\right)^{5/2}$$

i.e.

$$\frac{\dot{Q}}{\text{litre/s}} = \frac{0.84}{1000} \times \frac{10^3}{60} \times \frac{1}{316}\left(\frac{H}{\text{mm}}\right)^{5/2}$$

$$= 44.3 \times 10^{-6}\left(\frac{H}{\text{mm}}\right)^{5/2}$$

(c) Hence the empirical formula in the changed units may be written

$$v = 44.3 \times 10^{-6} z^{5/2}$$

with the qualifying statement 'where $v$ represents the *number* of litres per second and $z$ the *number* of millimetres of head above the notch'.

Thus, if $z = 100$, then $v = 44.3 \times 10^{-6} \times 10^5 = 4.43$; i.e. $\dot{Q} = 4.43$ litres/s.

**Example 31**

Deduce a numerical formula for the indicated power $P_i$ of a reciprocating-piston engine in kW and in which the mean effective pressure $P$ is in

$N/m^2$, the stroke $L$ in mm, the piston area $A$ in $mm^2$ and the cyclic frequency $C$ per min.

Indicated power is the rate of work done on the face of a piston (or pistons) in the cylinder(s) of an engine, and the mean effective pressure (i.e. per cycle) is usually estimated from indicator diagrams.

The physical formula for indicated power of such an engine is therefore

$$P_i = P\,L\,A\,C$$

which is independent of any particular system of units. Thus, imposing the particular units stated, we may write the physical equation as

$$\frac{P_i}{\text{kW}} = \frac{P}{\text{N/mm}^2} \times \frac{L}{\text{mm}} \times \frac{A}{\text{mm}^2} \times \frac{C}{1/\text{min}} \left\{\frac{\text{N mm}}{\text{kW min}}\right\}\left[\frac{\text{W s}}{\text{N mm} \times 10^6}\right]\left[\frac{\text{min}}{60\ \text{s}}\right]$$

or as the numerical formula

$$i = \frac{p\,l\,a\,c}{60 \times 10^6}$$

where $i$ is the *number* of kilowatts of indicated power, $p$ is the number of $N/mm^2$ on the piston, $l$ and $a$ are the numbers of mm and $mm^2$, respectively, and $c$ is the number of cycles per minute $= n$ or $n/2$ in two-stroke or four-stroke engines, respectively, where $n$ is the number of revolutions of the crank/min.

**Example 32**

Find the mechanical and brake thermal efficiencies, and draw up an energy account or balance sheet for a single-cylinder, four-stroke internal-combustion engine of stroke 203 mm and cylinder diameter 127 mm which on test gave the following data: fuel-oil of calorific value 41.8 MJ/kg consumed at the rate of 1.34 kg/h, crankshaft speed or rotational frequency 420 rev/min and brake torque 120 N m, indicated mean effective pressure 0.675 $N/mm^2$, cylinder jacket cooling water 81.8 kg/h with a temperature rise of 39 K, exhaust heat-exchanger water 84.5 kg/h with a temperature rise of 43.3 K. The specific heat-capacity of water is 4.19 kJ/kg K.

Using the numerical formulae of Examples 27 and 31 for brake and indicated powers, we get

$$b = \frac{2\pi t n}{60\,000} = \frac{2\pi \times 120 \times 420}{60\,000} = 5.26$$

i.e. brake power $P_b = 5.26$ kW and

$$i = \frac{plac}{60 \times 10^6} = \frac{0.675 \times 203}{60 \times 10^6} \times \frac{\pi}{4} (127)^2 \times 210 = 6.08$$

i.e. indicated power $P_i = 6.08$ kW.

Energy of combustion supplied in fuel

$$= \frac{1.34 \text{ kg} \times 41.8 \text{ MJ/kg}}{3.6 \text{ ks}} \left[\frac{\text{kW s}}{\text{kJ}}\right] = 15.6 \text{ kW.}$$

Hence

$$\text{mechanical efficiency} = \frac{5.26}{6.08} = 0.865 \text{ or } 86.5\%$$

$$\text{brake thermal efficiency} = \frac{5.26}{15.6} = 0.336 \text{ or } 33.6\%$$

Rate of heat-transfer to cylinder jacket water

$$= \frac{81.8 \text{ kg}}{3.6 \text{ ks}} \times \frac{4.19 \text{ kJ}}{\text{kg K}} \times 39 \text{ K} \left[\frac{\text{kW s}}{\text{kJ}}\right] = 3.71 \text{ kW}$$

Rate of heat-transfer in exhaust heat-exchanger

$$= \frac{84.5 \text{ kg}}{3.6 \text{ ks}} \times \frac{4.19 \text{ kJ}}{\text{kg K}} \times 43.3 \text{ K} = 4.26 \text{ kW}$$

Thus, referring to *Figure 4.1* (and assuming that the conditions are steady), by the law of conservation of energy the rates of inflow and outflow of energy across the boundary enclosing the system (in this case, an energy transformer) are the same, i.e.

$$F + A + J_i + W_i + S_i = P_b + E + J_o + W_o + S_o$$

where each symbol represents an amount of energy per unit time. This physical equation (an identity) can be rearranged as

$$F = P_b + (J_o - J_i) + (W_o - W_i) + \{(E - A) + (S_o - S_i)\}$$

which provides the basis for the following energy balance sheet.

| *Energy stream* | kW | % |
|---|---|---|
| Energy supplied in fuel ($F$) | 15.60 | 100.0 |
| Energy delivered as brake power ($P_b$) | 5.26 | 33.6 |
| Energy transferred to jacket water ($J_o - J_i$) | 3.71 | 23.8 |
| Energy transferred in heat-exchanger ($W_o - W_i$) | 4.26 | 27.3 |
| Energy in exhaust and to radiation (by difference) | 2.37 | 15.3 |
| Totals | 15.60 | 100.0 |

**Example 33**

Deduce physical and numerical formulae for the velocity of gas flowing adiabatically through a horizontal nozzle from a large vessel which makes the velocity of approach negligible.

*Physical formula*

From the steady-flow energy equation for *unit mass*, namely:

$$Q - W = \Delta\left(\frac{p}{\rho} + \frac{u^2}{2} + zg + c_v T\right)$$

we deduce that when work $W = 0$ and $Q = 0$ (adiabatic) and $z = 0$ (horizontal) the physical equation

$$\frac{p}{\rho} + \frac{u^2}{2} + c_v T = \text{constant}$$

applies. Hence, since $p = \rho R T$ for a gas, we deduce

$$\frac{u^2}{2} + T(c_v + R) = \frac{u^2}{2} + c_p T = \text{constant}$$

as the energy equation per unit mass of gas flowing adiabatically (and *not* assumed reversibly, i.e. isentropically) through the nozzle.

Alternatively, $c_p dT + d(\frac{1}{2}u^2) = 0 = dh + d(\frac{1}{2}u^2)$ since, by definition, specific enthalpy $h = e + pv = e + RT$ for a gas, and hence $dh = (c_v + R)\, dT = c_p dT$.

Thus, from the integral $\int_1^2 d(\frac{1}{2}u^2) = -\int_1^2 dh$, we get

$$\tfrac{1}{2}(u_2^2 - u_1^2) = h_1 - h_2 = \text{the decrease of specific enthalpy}$$

and, if $u_1$ is negligible,

$$u_2 = \sqrt{\{2(h_1 - h_2)\}} = \sqrt{\{2c_p(T_1 - T_2)\}}$$

which is the physical formula required.

*Numerical formula*

If it is decided to have a formula in which velocity is in m/s and enthalpy in kJ/kg, the *physical formula* can be rewritten as

$$\left(\frac{u^2}{\text{m/s}}\right)^2 = 2\left(\frac{h_1 - h_2}{\text{kJ/kg}}\right)\left\{\frac{\text{s}^2}{m^2} \times \frac{\text{kJ}}{\text{kg}}\right\}\left[\frac{\text{kg m}}{\text{N s}^2}\right]\left[\frac{\text{kN m}}{\text{kJ}}\right]$$

$$= 2 \times 10^3 \left(\frac{h_1 - h_2}{\text{kJ/kg}}\right)$$

Hence, using the same letter symbols (as is often done, and is therefore sometimes the cause of confusion because of ambiguity in the use of letter symbols; see page 62), the corresponding *numerical formula* may be written

$$u_2 = \sqrt{\{2000(h_1 - h_2)\}} = 44.7\sqrt{(h_1 - h_2)}$$

which is incomplete unless qualified by the statement 'where $u_2$ is the number of m/s and $(h_1 - h_2)$ is the reduction in enthalpy in kJ/kg'. Thus $u_2 = 44.7\sqrt{(h_1 - h_2)}$ is a restricted formula attached to particular units, as are all numerical and empirical formulae.

## Physical formulae and insertion of units

Although the choice of letter symbols to denote physical quantities does not imply the choice of particular units, each letter symbol must be replaced (in quantitative work) by a number and its unit if reduction by cancellation of like units is to be free from errors with certainty, especially when products and quotients of several physical quantities are involved. Hence, if an answer is seen to have worked out with the wrong dimensions (e.g. a pressure working out to a unit of, say, N/K instead of $N/m^2$), this reveals that a mistake has been made and will induce a check to be made on the analysis and/or substitutions of numbers × units for physical quantities together with the multiplying unities or unity brackets used in the reduction process; if they have been written down, the check is so easy to do.

Thus the numbers and their units safeguard each other, and *it is always desirable to write in all the units involved rather than risk errors by trying to carry them along in one's head.*

## Example 34

Show that the insertion of units in the working of problems prevents errors and can induce a check to be made on analysis and quantitative work.

The letter symbol $R$ is often used to represent the gas constant generally, i.e. the universal molar gas constant and the characteristic (or specific) constant of a gas or gas-mixture (see Examples 19 and 22, also Appendix 5).

If, in the characteristic gas equation $PV/T = Rm$, the universal (molar) gas constant ($R$ = 8.3143 J/mol K) is substituted instead of the characteristic (specific) gas constant, say, for air ($R$ = 287 J/kg K), then the amount of substance would, necessarily, have to be the number of moles (mol) instead of the mass in kilograms (kg), otherwise the units would not reduce to those of $PV/T$, namely

$$\frac{\text{N m}}{\text{K}} \quad \text{or} \quad \frac{\text{J}}{\text{K}}$$

but would be $\quad \dfrac{\text{J}}{\text{mol}} \dfrac{\text{kg}}{\text{K}}$

and thus would require the ratio kg/mol to be reduced by the unity multiplier for air, namely

$$\left[\frac{\text{mol}}{0.029\ \text{kg}}\right]_{\text{air}}$$

Alternatively, the ideal gas equation $PV = RnT$ could be used, in which $n = m/M$, where $n$ is the amount of gas in moles and $M$ is the molar mass.

When the double equation, $PV = RmT = \overline{R}nT$, is written it is then, however, desirable to distinguish between the '$R$'s for the characteristic (specific) and universal (molar) gas constants in order to be able to be quite clear about

$$Rm = \overline{R}n = \overline{R}\frac{m}{M}$$

Hence $\quad R = \overline{R}/M$

which, for air, is

$$\frac{8.3143\ \text{J/mol K}}{0.029\ \text{kg/mol}} = 287\ \text{J/kg K}$$

Also, for air $c_p = 1.005$ kJ/kg K; hence, since $c_p - c_v = R$, $c_v =$ 0.718 kJ/kg K for air.

### Different algebras

Although the foregoing statements and examples indicate different ways in which letter symbols can be used, it is worth reiterating, because of the need of being constantly aware that, in addition to representing numbers or physical quantities, letter symbols may also be used to represent vector quantities. Thus, in analysis, the same letter symbols are often used in different algebras, namely:

($|x|$) *numerical algebra*, which is generalised arithmetic. For example,

$$\text{number } |u| = 5$$

$$\sin\theta = \theta - \frac{\theta^3}{6} + \frac{\theta^5}{120} \ldots$$

$$\epsilon^x = 1 + x + \frac{x^2}{2} + \frac{x^3}{6} + \ldots$$

where $\theta$ and $x$ are numbers.

($x$) *quantity algebra,* in which symbols represent physical quantities in accordance with the Stroud convention. For example,

$$\text{speed } u = \frac{\mathrm{d}s}{\mathrm{d}t} = \omega r = 5 \text{ m/s}$$

($\bar{x}$) *vector algebra*, in which symbols represent magnitude (i.e. number × unit), direction and sense. For example

$$\text{velocity } u = 5 \text{ m/s at } 48.6^\circ \text{ to the abscissa}$$

or $$\bar{u} = 3\bar{i} + 4\bar{j}$$

where $\bar{i}$ and $\bar{j}$ are unit vectors (of magnitude 1 m/s) along Cartesian coordinates.

Hence it should be realised:

(a) that, although the dimensions of, say, torque and mechanical energy are the same, i.e. $[\mathrm{ML^2/T^2}]$, an essential difference lies in the fact that they are the vector and scalar products, respectively, of the same two physical (vector) quantities: force and distance ($\bar{F} \times \bar{S}$ and $\bar{F} \cdot \bar{S}$);

(b) *that the same dimension does not necessarily imply the same kind of physical quantity, as is the case vice versa.* For example: (i) torque and energy; (ii) angular momentum has the same dimensions, $[\mathrm{ML^2T^{-1}}]$, as dynamic viscosity × volume, but there is clearly no physical connection.

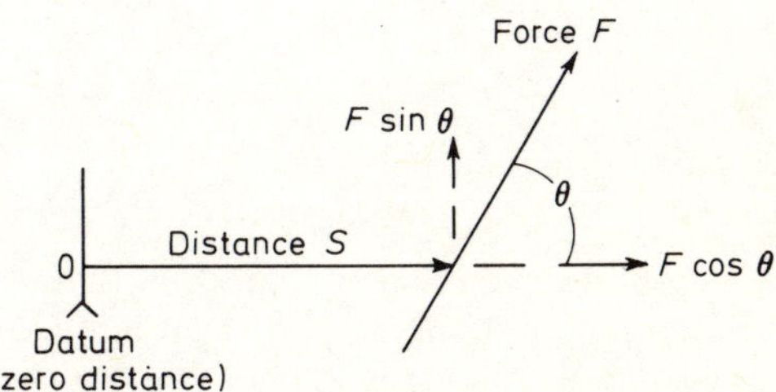


*Figure 4.2 Scalar and vector products*

Referring to *Figure 4.2*, we may note that the scalar (or dot) product of two vectors $\bar{F}$ and $\bar{S}$ of magnitude (number × unit) $f$ and $s$, respectively, and whose lines of action are inclined at angle $\theta$, is

$$\bar{F} \cdot \bar{S} = fs \cos\theta = E$$

and their vector (or cross) product is

$$\overline{F} \times \overline{S} = |fs \sin\theta|\,\overline{p} = \overline{T}$$

where $\overline{p}$ is the unit vector (say 1 N m) which gives the direction of $\overline{T} = \overline{F} \times \overline{S}$, i.e. perpendicular to the plane containing $\overline{F}$ and $\overline{S}$.

5

# Graphical differentiation and integration

## Manipulation of scales

There are relatively few problems in engineering and applied science in which the relationships between the variables can be represented by exact mathematical laws. In general, therefore, when analysing the results of experiments it is necessary to draw graphs of direct observations or measurements from which further deductions can be made by approximate methods. Some of these methods (e.g. Simpson's rules) are well known but, because of scales of ordinates and abscissae of the graphs, difficulties often arise in deducing the correct numerical answers. Difficulties with scales are easily overcome if all letter symbols in analysis represent physical quantities, i.e. if each is composed of a number × a unit. This can be shown by considering the simple case of the kinematics of a moving body.

From a distance-time $(S, t)$ graph (*Figure 5.1*) the first derivative $(\mathrm{d}S/\mathrm{d}t = u)$ gives a speed-time $(u, t)$ graph (*Figure 5.2*), and the second derivative $(\mathrm{d}^2S/\mathrm{d}t^2 = \mathrm{d}u/\mathrm{d}t = a)$ gives an acceleration-time $(a, t)$ graph.

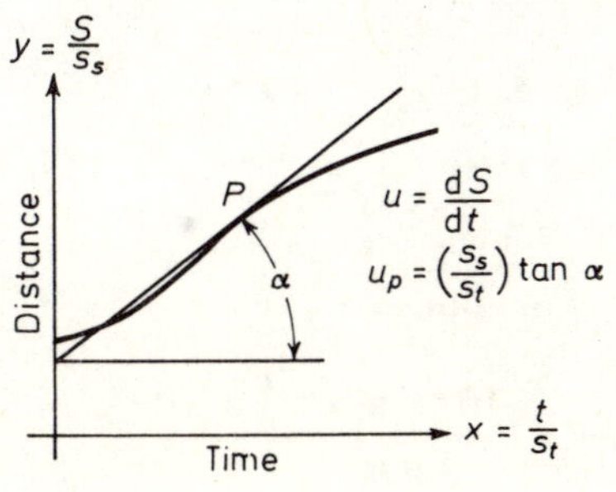


*Figure 5.1 Distance–time graph*

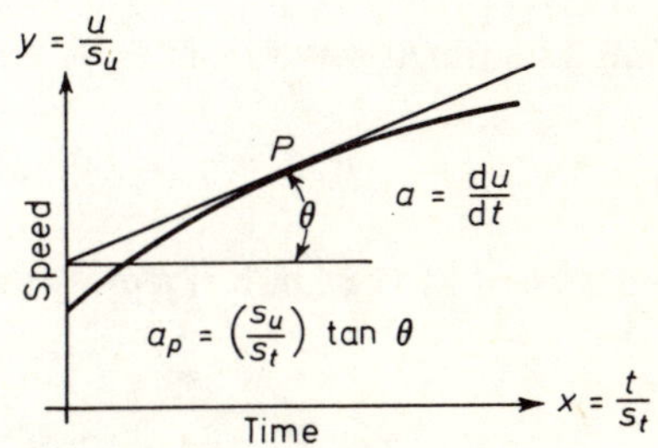


*Figure 5.2 Speed–time graph*

In *Figure 5.2*, tan $\theta$ (the slope at any point P) represents the acceleration at the corresponding instant of time, and the area under the curve represents the distance moved in a corresponding interval of time. Similarly, in the case of the speed–distance graph of *Figure 5.3*, the acceleration at any point P is represented by the length of the sub-normal (see Example 36), but difficulty is sometimes experienced in finding the scale to which the length of the subnormal represents acceleration.

**Example 35**

Use the speed–time graph of *Figure 5.2* to show the involvement of scales of speed and time axes on the acceleration and distance covered. Also, estimate the numerical values of acceleration and distance covered in the cases given.

Assuming that the speed–time curve is plotted on millimetre squared paper, areas will conveniently be in mm$^2$.

The scale of speed is defined as speed ÷ length of ordinate,

i.e. $$s_u = \frac{u}{y} = \frac{3 \text{ m/s}}{10 \text{ mm}} \text{ (say)}$$

Similarly, the scale of time is time ÷ length of abscissa,

i.e. $$s_t = \frac{t}{x} = \frac{1 \text{ s}}{2 \text{ mm}} \text{ (say)}$$

Alternatively, the speed = ordinate × scale of speed, or

$$u = y\, s_u$$

and the time = abscissa × scale of time, or

$$t = x \times s_t$$

Thus an ordinate $y$ = 60 mm represents a speed

$$u = y\, s_u = 60 \text{ mm} \times \frac{3 \text{ m/s}}{10 \text{ mm}} = 18 \text{ m/s}$$

and a speed of 108 km/h is represented by an ordinate

$$y = \frac{\text{speed}}{\text{scale of speed}} = \frac{u}{s_u} = \frac{108 \text{ km/h}}{\dfrac{3 \text{ m/s}}{10 \text{ mm}}}$$

$$= \frac{108 \times 10^3 \text{m}}{3600 \text{ s}} \times \frac{10 \text{ mm s}}{3 \text{ m}} = 100 \text{ mm}$$

The acceleration at any point P on *Figure 5.2* is

$$a = \frac{du}{dt} = \frac{d(y\, s_u)}{d(x\, s_t)} = \left(\frac{s_u}{s_t}\right) \frac{dy}{dx} = \frac{s_u}{s_t} \tan\theta$$

Thus, if the slope of the tangent at point P is $\tan\theta = 0.25$, the acceleration at that point is

$$a = \left(\frac{3 \text{ m}}{10 \text{ mm s}}\right) \left(\frac{2 \text{ mm}}{1 \text{ s}}\right) \times \tfrac{1}{4} = 0.15 \text{ m/s}^2$$

The distance covered during an interval of time $t_1$ to $t_2$ can be found from *Figure 5.2* as follows. Since $u = dS/dt$, then

$$S_2 - S_1 = \int_{t_1}^{t_2} u \, dt = \int_{x_1}^{x_2} (y\, s_u)\, d(x\, s_t) = (s_u\, s_t) \int_{x_1}^{x_2} y \, dx$$

i.e. distance covered = $(s_u\, s_t)$ (area under curve between $x_1$ and $x_2$)

Thus, if the area under the curve between $x_1$ and $x_2$ (i.e. during an interval of time $(t_2 - t_1)$) is 500 mm$^2$, the distance covered in the time $(t_2 - t_1)$ is

$$\left(\frac{3 \text{ m}}{10 \text{ mm s}}\right) \left(\frac{2 \text{ mm}}{1 \text{ s}}\right) \times 500 \text{ mm}^2 = 75 \text{ m}$$

**Example 36**

(a) Prove that the length of the sub-normal of a particular point on a speed–distance graph is a measure of the acceleration of a moving body at that point.

(b) Estimate the acceleration at a point where the length of the sub-normal is 84 mm on a speed–distance graph whose ordinate scale is

$$s_u = \frac{1 \text{ m/s}}{1 \text{ mm}}$$

and abscissa scale is

$$s_s = \frac{60 \text{ m}}{1 \text{ mm}}$$

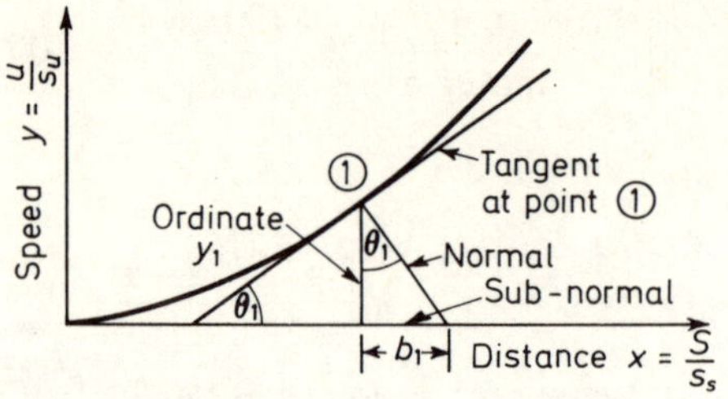


*Figure 5.3 Speed–distance graph*

(a) Let the speed–distance graph be plotted as in *Figure 5.3* and the tangent, normal and sub-normal be drawn for any particular point ①. Then, since the acceleration of a body is given by

$$a = \frac{\mathrm{d}u}{\mathrm{d}t} = \frac{\mathrm{d}u}{\mathrm{d}S} \times \frac{\mathrm{d}S}{\mathrm{d}t} = u\,\frac{\mathrm{d}u}{\mathrm{d}S}$$

the acceleration at the particular point ① is

$$a_1 = u_1 \left(\frac{\mathrm{d}u}{\mathrm{d}S}\right)_1 = (s_u\, y_1) \left\{\frac{\mathrm{d}(s_u y)}{\mathrm{d}(s_s x)}\right\}_1 = \frac{s_u^{\;2}}{s_s}\, y_1 \left(\frac{\mathrm{d}y}{\mathrm{d}x}\right)_1$$

Hence

$$a_1 = \frac{s_u^{\;2}}{s_s}\, y_1 \tan\theta_1 = \frac{s_u^{\;2}}{s_s}\, y_1 \left(\frac{b_1}{y_1}\right) = \frac{s_u^{\;2}}{s_s}\, b_1$$

i.e. the acceleration ($a$) of the body is proportional to the length ($b$) of the sub-normal.

(b) In particular, if the sub-normal for point ① is $b_1 = 84$ mm, then the acceleration of the body at that point is

$$a_1 = \frac{s_u^{\;2}}{s_s}\, b_1 = \left(\frac{1\ \mathrm{m}}{1\ \mathrm{mm\ s}}\right)^2 \left(\frac{1\ \mathrm{mm}}{60\ \mathrm{m}}\right) \times 84\ \mathrm{mm} = 1.4\ \mathrm{m/s^2}$$

**Bending and deflection of beams**

In order to find numerical answers to problems involving shearing force, bending moment, slope and deflection diagrams, areas under graphs plotted to particular scales are often required. For example, since a shearing-force ($F,L$) diagram can usually be drawn (*Figure 5.4*), bending moment at any section of the beam ($M = \int F\mathrm{d}L$) is equal to the area

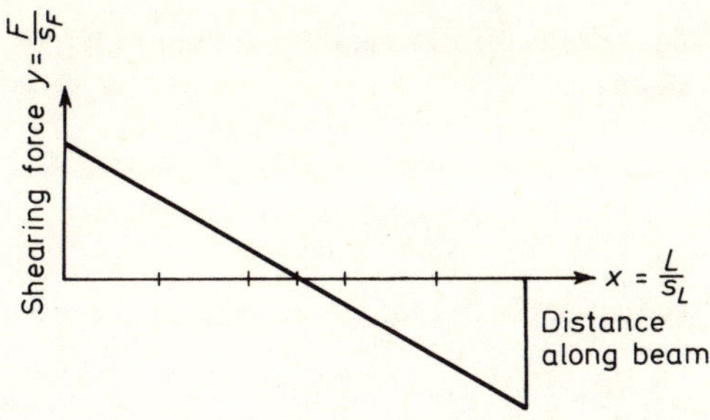


*Figure 5.4 Shearing-force graph plotted to scales $s_F$ and $s_L$*

under the shearing-force line up to that section multiplied by the appropriate scales.

**Example 37**

If the scales of the shearing-force diagram are $s_F = 3.5$ N/mm and $s_L = 1$ m/40 mm or 25 and the weight per unit length of a uniformly-loaded, simply-supported beam is $w = 45$ N/m on a span of 3 m, the shearing-force diagram plotted to these scales will be as shown in *Figure 5.4*, and the bending moment at, say, the centre of the beam can be determined quantitatively from

$$M_{L/2} = \int_0^{L/2} F\mathrm{d}L = \int_0^{L/2} (s_F y)\,\mathrm{d}\,(s_L x) = (s_F s_L)\int_0^{L/2} y\,\mathrm{d}x$$

$$= (s_F s_L) \times \text{ area under } x, y \text{ graph up to mid-span}$$

i.e. $$M_{L/2} = \left(3.5\,\frac{\mathrm{N}}{\mathrm{mm}} \times 25\right) \times \tfrac{1}{2}\,(19.3 \text{ mm} \times 60 \text{ mm})$$

$$= 51 \text{ N m}$$

This result can easily be checked by taking moments in this simple case, but in more complicated cases the graphical method must be resorted to.

The slope $(\mathrm{d}\delta/\mathrm{d}L)$ and deflection $(\delta)$ of beams can be estimated in a similar way by graphical integration based on the relationship

$$\frac{\mathrm{d}^2\delta}{\mathrm{d}L^2} = \frac{M}{EI}$$

i.e. deflection ($\delta$) is derived from bending moment ($M$) by double integration or, since

$$\frac{d^4\delta}{dL^4} = \frac{w}{EI}$$

from the loading diagram by quadruple integration.

# Appendix 1

**Definitions of the seven base units of the SI**

*Metre.* The metre is the length equal to 1 650 763.73 wavelengths in vacuum of the radiation corresponding to the transition between the levels $2p_{10}$ and $5d_5$ of the krypton-86 atom.
*Kilogram.* The kilogram is the unit of mass; it is equal to the mass of the international prototype of the kilogram.
*Second.* The second is the duration of 9 192 631 770 periods of the radiation corresponding to the transition between the two hyperfine levels of the ground state of caesium-133 atom.
*Ampere.* The ampere is that constant current which, if maintained in two straight parallel conductors of infinite length, of negligible circular cross-section, and placed 1 metre apart in vacuum, would produce between these conductors a force equal to $2 \times 10^{-7}$ newton per metre of length.
*Kelvin.* The kelvin, unit of thermodynamic temperature, is the fraction 1/273.16 of the thermodynamic temperature of the triple point of water. See Appendix 3.
*Candela.* The candela is the luminous intensity, in the perpendicular direction, of a surface of 1/600 000 square metre of a black body at the temperature of freezing platinum under a pressure of 101 325 newtons per square metre (i.e. the standard atmosphere).
*Mole.* The mole is the amount of substance of a system which contains as many elementary entities as there are atoms in 0.012 kilogram of carbon-12. The elementary entity must be specified and may be an atom, a molecule, an ion, an electron, a photon, etc., or a specified group of such entities (e.g. 24 g of C $\equiv$ 2 mol of C). See Appendix 5.

**Definitions of the two supplementary units**

*Radian.* The radian is the angle between two radii of a circle which cut off on the circumference an arc equal in length to the radius (e.g. the

angle of a semi-circle is $\pi$ rad). See Appendix 2.
*Steradian.* The steradian is the solid angle which, having its vertex in the centre of a sphere, cuts off an area of the surface of the sphere equal to that of a square having sides of length equal to the radius of the sphere (e.g. the solid angle of a hemisphere is $2\pi r^2/r^2 = 2\pi$ sr).

# Appendix 2

## Plane angle: an exception to the Stroud convention

All the usual physical formulae involving plane angle and angular motion are 'restricted' in that they are derived on the presupposition that circular (i.e. radian) measurement is used in quantifying plane angles, as is the case in mathematics generally.

*Plane angles* are usually expressed in degrees, radians, revolutions or turns. Sexagesimal measure, in which a circle or one revolution is divided into 360 degrees, may be thought of in terms of sectors between straight lines radiating from a point or apex, as on a protractor. Circular measure, however, expresses the size of any angle by the ratio of the circular arc bounded by it to the radius of the circle, the radius being of any convenient length, and its centre being the apex of the angle.

In the SI the radian is defined as 'the angle between two radii of a circle which cut off on the circumference an arc equal in length to the radius (e.g. the angle of a semi-circle is $\pi$ rad)'. Alternatively, the radian is that angle subtended at the centre of a circle by an arc the length of which is equal to the radius of the circle.

Numerical factors interrelate the customary modes of expressing plane angle, namely

$$1 \text{ rad} = \frac{180 \text{ deg}}{\pi} = \frac{1 \text{ rev}}{2\pi}$$

Any plane angle may be expressed as $\theta = |\theta_1| \text{ rad} = |\theta_2| \text{ deg}$ in which the ratio of the numerics

$$\frac{|\theta_2|}{|\theta_1|} = \frac{180}{\pi} = 57.3 = \frac{\text{rad}}{\text{deg}}$$

Although, in general, units only enter into a problem towards the end when actual sizes of physical quantities are involved, an exception to this is plane angle ($\theta$ and its derivatives $\dot{\theta}$ or $\omega$ and $\ddot{\theta}$ or $\alpha$). From the outset of mathematical analysis, plane angle is presupposed to be in circular (i.e. radian) measure, and thus is an exception (the only one) to the Stroud convention. Otherwise the symbology would be much more cumbersome, and the calculus less elegant. For example if, say, 'radr' were used to indicate an arc/radius ratio, analogous to the way in which 'sin' is used to indicate the opposite side/hypotenuse ratio of a right-angled triangle, this would render the 'radian ratio' (radr $\theta$) of any angle $\theta$ independent of particular units. Thus radr $\theta$ would be a pure number as are sin $\theta$, cos $\theta$, etc; and radr $\theta$ = arc $x$/radius $r$ = $|\theta| = (\theta/\text{rad})$ would be alternatives. In particular, if $x = r$, $\theta = 1$ rad and radr (1 rad) = 1, the latter being a shorthand way of defining radian as a unit of plane angle. However, 'radr' would necessarily recur throughout the symbology and would tend to clutter up the analytical work and formulae involving plane angle and its derivatives until it could be eliminated by use of the 'unity bracket' [radr (1 rad)] = 1. For example, instead of the customary formulae

$$W = T\theta; \; v = r\omega; \; T = I\alpha$$

we would have to write

$$W = T(\text{radr}\,\theta); \; v = r(\text{radr}\,\omega)$$

$$\begin{aligned} T = I(\text{radr}\,\alpha) &= 9\ \text{kg m}^2 \times \text{radr}\,(60\ \text{deg/s}^2) \text{ in a particular case} \\ &= 9\ \text{kg m}^2 \times \text{radr}\left(\frac{\pi}{3}\frac{\text{rad}}{\text{s}^2}\right) \\ &= 3\pi\ \text{kg}\left(\frac{\text{m}}{\text{s}}\right)^2 [\text{radr}\,(1\ \text{rad})] = 3\pi\ \text{N m} \end{aligned}$$

Such 'radr' encumbrances can be (and are) avoided by presupposing circular (i.e. radian) measurement of plane angle in the mathematics. This is the reason for the tacit assumption that radian is the mode of measurement of the plane angles involved in quantifying or expressing plane angle ($\theta$), angle rate ($\omega$) and angular acceleration ($\alpha$), i.e. it is the *numbers* ($|\theta|$, $|\omega|$, $|\alpha|$) of radians that are required in quantitative work involving $\theta$, $\omega$ and $\alpha$. For example, in the simple case of a particle of mass $m$ rotating about a fixed point at a radius $r$ with angular velocity $\omega$, its instantaneous linear (or tangential) velocity is given by $\dot{x} = v = r\omega$ *only* if $\omega$ represents the *number* of radians per any unit of time. This can be emphasised by writing

$$v = r\,|\omega| = r\left(\frac{\omega}{\text{rad}}\right)$$

The same restriction is implicit in $E = \frac{1}{2}I\omega^2$ for kinetic energy of the rotating particle, made explicit by writing

$$E = \tfrac{1}{2}I\left(\frac{\omega}{\text{rad}}\right)^2 = \tfrac{1}{2}I(\,|\omega|\,)^2$$

In fact, all the customary physical formulae involving plane angle and its derivatives are 'restricted' in one particular, namely that they contravene the Stroud convention by implying circular or the radian mode of measurement of plane angles. Thus, if angles are first measured or expressed in degrees, revolutions or turns, they must at some stage in quantitative work be converted into radians so that the *correct number* is used.

It may be contended that circular measure of plane angle (being a ratio of like quantities) is a number, i.e. is dimensionless (e.g. the circular or radian measure of the plane angle of a semi-circle is $\pi$). Hence it may be said that radian should not appear as a unit, e.g. angular velocity $\omega$ and angular acceleration $\alpha$ should have units 1/s and 1/s$^2$, respectively. It is, however, helpful in quantitative work to insert 'rad' as an indicator (if only to make sure that plane angles have finally been quantified by circular measurement required by the customary formulae involving them), and then to apply the simple artifice of replacing rad by unity at any convenient point (usually at the end, as in Examples 12, 14, 15, 27, 28) in the quantitative or arithmetical stage of a problem. This procedure not only yields correct answers (as proved by laboratory and full-scale tests), but is irreducibly simple and also enables the Stroud convention of replacing each letter symbol by its value (number × unit) to be adhered to in respect of angles in 'restricted' physical formulae, either (a) with or (b) without rad as the divisor of plane angle and its derivatives. For example, torque is

$$T = I\left(\frac{\alpha}{\text{rad}}\right) = 10\ \text{kg m}^2\left(\frac{2}{\text{rad}}\,\frac{\text{rad}}{\text{s}^2}\right)\left[\frac{\text{N s}^2}{\text{kg m}}\right] = 20\ \text{N m} \quad \text{(1a)}$$

$$T = I\alpha = 10\ \text{kg m}^2\left(2\,\frac{\text{rad}}{\text{s}^2}\right)\left[\frac{1}{\text{rad}}\right] = 20\ \text{N m} \quad \text{(1b)}$$

and couple in a gyroscope is

$$C = I\left(\frac{\Omega}{\text{rad}}\right)\left(\frac{\omega}{\text{rad}}\right) \quad \text{(2a)}$$

$$= 0.3\ \text{N s}^2\ \text{mm}\left(\frac{300}{\text{rad}}\,\frac{\text{rev}}{\text{s}}\right)\left(\frac{0.02}{\text{rad}}\,\frac{\text{deg}}{\text{s}}\right)$$

$$= 1.8\ \text{N mm}\,(2\pi)\left(\frac{\pi}{180}\right) = \frac{\pi^2}{50}\ \text{N mm}$$

$$C = I\Omega\omega \tag{2b}$$

$$= 0.3 \text{ N s}^2 \text{ mm} \left(300 \frac{\text{rev}}{\text{s}}\right) \left(0.02 \frac{\text{deg}}{\text{s}}\right) \left[\frac{2\pi}{\text{rev}}\right] \left[\frac{\pi}{180 \text{ deg}}\right]$$

$$= \frac{\pi^2}{50} \text{ N mm}$$

As expected, both methods, (a) and (b), give the correct result, and of these (b) is the least cumbersome and most straightforward.

# Appendix 3

## Temperature

Two separate concepts, thermodynamic and thermometric, need to be appreciated in connection with temperature and its symbology. Thermodynamic temperature is based on a corollary of the second law of thermodynamics and formulated in the simple equation $T_1/T_2 = (Q_1/Q_2)_{\text{reversible}}$ while thermometric temperature is based on such properties as the expansion of particular substances, e.g. mercury.

Temperature is an intensive property of a system (i.e. of any collection of matter within a prescribed boundary), and temperature-difference is the cause of the escaping tendency of energy by the process of heat-transfer.

*Thermodynamic temperature,* independent of any particular substance or temperature scale, is designated by the letter symbol $T$ and measured by the SI unit named 'kelvin', unit symbol K. This unit is also used to express thermodynamic temperature-difference (in a similar way to pressure and pressure-difference which have the same unit, $N/m^2$), context implying whether it is a temperature-difference or a temperature above a zero (either absolute zero or conventional zero Celsius, abbreviated $0°C$, which is 273.15 K on the absolute scale).

Thus thermodynamic temperature is treated as any other basic physical quantity, the kelvin being defined as the fraction 1/273.16 of the thermodynamic temperature of the triple point of ordinary water (i.e. saturation state where ice, water and steam are together in equilibrium). A thermodynamic temperature $T$ expressed in kelvins is commonly referred to as the *absolute* temperature in order to distinguish it from Celsius temperature.

*Celsius temperature,* designated by the letter symbol $t$, is defined as $t = T - T_0$, where $T_0 = 273.15$ K. Thus Celsius temperature is a truncated thermodynamic temperature, and the kelvin is the common unit, e.g. $t = c$ Celsius ($c$ °C) implies that $T = (c + 273.15)$ K.

*Practical temperature scales*

In practice, empirical scales devised for thermometers and other temperature-measuring devices usually register on the Celsius scale and, because of common usage of the term 'degrees Celsius', the abbreviation '°C' is attached to numbers indicative of Celsius temperature levels. The International Practical (Kelvin-Celsius) Temperature Scale (IPTS, 1968) embodies the most accurate values currently available for thermodynamic temperatures of a number of fixed points (e.g. triple point of hydrogen and freezing point of gold) in addition to the arbitrary but exact 273.16 K and 373.15 K (or 0.01 °C and 100 °C) for the triple and boiling points of water at 1 standard atmosphere, respectively.

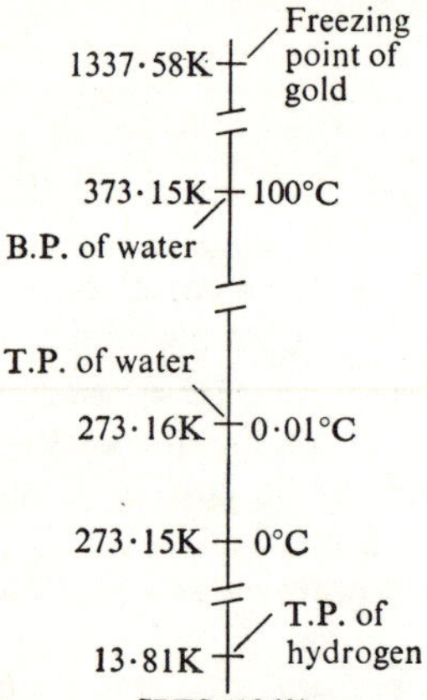

1337·58K — Freezing point of gold

373·15K — 100°C — B.P. of water

T.P. of water — 273·16K — 0·01°C

273·15K — 0°C

13·81K — T.P. of hydrogen

IPTS (1968)

Thus IP Celsius scale-temperatures can be taken as being equal to thermodynamic temperatures in kelvins less 273.15. The IPTS is currently the best practical tool for measuring thermodynamic temperature, and instruments used in science and technology are calibrated in terms of it. The conventional abbreviation '°C' is *not* a unit symbol but is a scale-indicator (or 'gauge' temperature, analogous to gauge pressure) denoting that Celsius temperature is being quoted; a unit-interval of Celsius scale temperature is K, the SI unit of thermodynamic temperature, i.e. K is the unit symbol for temperature difference whether absolute or Celsius. For example, the interval between temperature $t_1 = a_1$ Celsius ($a_1$ °C) and $t_2 = a_2$ Celsius ($a_2$ °C) is $t_1 - t_2 = (a_1 - a_2)$ K, not $(a_1 - a_2)$ °C, for to use the abbreviation '°C' as a unit of temperature as well as K would contravene an aim of the SI, namely, of using one unit symbol for the same unit quantity, and it would also confuse with the use of '°C' as a scale-indicator. Thus a temperature formerly specified as $T = b$ °K should, in the SI, be specified as $T = b$ K, and $t = a$ °C implies a thermodynamic temperature $T = (a + 273.15)$ K.

# Appendix 4

## Frequency

The word 'frequency' is associated with periodic phenomena and the recurrence of similar events, e.g. the number of repetitions of a regular occurrence per unit of time. Frequency is the inverse of periodic time, and this is probably the reason why the special name 'hertz' has been listed by authoritative committees of the SI as a derived unit (Hz = 1/s) and defined as 'the frequency of a periodic phenomenon of which the period is one second'; this can be ambiguous unless there is an understanding (implied if not actually stated) of what recurs each second. For example, a drop-forge hammer striking 60 blows a minute may be said to have an impact frequency of 1 Hz, where Hz means blows/s, and (if preferred!) the motion of the earth relative to the sun a rotational frequency of 3.17 nHz instead of 1/annum. Hence, just as the current authoritative list of special names of the SI reveals that, by definition,

$$\left[\frac{\mathrm{N\ s^2}}{\mathrm{kg\ m}}\right] = 1, \quad \left[\frac{\mathrm{W\ s}}{\mathrm{N\ m}}\right] = 1, \text{etc.}$$

it would at first appear that [Hz s] = 1 for all frequencies without qualification. This is not so, however, because in engineering and elsewhere the word 'frequency' can be interpreted (and is customarily used) in different ways according to the phenomena involved. For example, for rotational, angular and cyclic rates the customary modes of measurement are rev/s, rad/s and cycle/s, respectively, linked by a numerical factor of $2\pi$ in the case of rev and rad:

$$\left[\frac{2\pi\ \mathrm{rad}}{1\ \mathrm{rev}}\right] = 1 \quad \textit{always}$$

Thus we may express the frequency (say $q$) of a rotating vector or crank representative of *one cycle each revolution* (e.g. s.h.m. (see

Example 14), sine and cosine waves) as

$$q = |n| \text{ rev/s} = |\omega| \text{ rad/s} = |f| \text{ cycle/s}$$

where $|n|$, $|\omega|$ and $|f|$ are the *numerical* values of the *same* physical quantity, $q$, measured in different ways, of which only the measure of angular frequency in rad/s is 1/s (radian, being the ratio of equal lengths, is unity). Hence to designate a recurring physical quantity or phenomenon merely by the word 'frequency', of unit symbol Hz = 1/s, can be confusing; e.g. the ratio of cyclic to angular frequency is $2\pi$ for rotary and harmonic motions, and one cycle is performed in one revolution of the crank in two-stroke engines and in two revolutions in four-stroke engines (see page 17 and Example 31).

Committees promulgating the SI have not succeeded in giving 'frequency' an unambiguous or unique definition as a physical quantity or a description in terms of the unit hertz (because it can't be done!). Hence, *to avoid numerical errors caused by possible ambiguous use of hertz, if* Hz *is to be used at all, it must be given precise definition appropriate to the particular type of periodic phenomenon involved* (see page 17); in cases where there are several periodic phenomena or frequencies interrelated by numerical factors (e.g. 1 cycle/s = 1 rev/s = $2\pi$ rad/s) it must be clearly understood which of the frequencies is designated as 1 Hz. Thus electrical engineers now use Hz as meaning cycle/s (e.g. 1 kHz for 1 kilocycle/s), as do mechanical engineers for rotary, oscillatory and harmonic motions in which the usage of Hz is specified as 1 Hz = 1 rev/s = 1 cycle/s = $2\pi$ rad/s, i.e.

$$\left[\frac{\text{Hz s}}{2\pi}\right] = 1$$

is the unity multiplier for conversion of units in such cases (see Examples 14 and 15).

Alternatively, the letter symbol $q$ may be used to represent the magnitude of cyclic frequency of vibrational or oscillatory motion, i.e. number of cycles per unit time; the corresponding periodic time $\tau$ is given by $\tau q = 1$ cycle. Consequently we may write

$$\frac{q}{2\pi \text{ rad/s}} = \frac{q}{\text{rev/s}} = \frac{q}{\text{cycle/s}} \quad \text{and} \quad \frac{\tau}{\text{s}} = \frac{\text{cycle/s}}{q}$$

$$\text{or} \quad \frac{q}{\text{rad/s}} = 2\pi\left(\frac{q}{\text{rev/s}}\right) = 2\pi\left(\frac{q}{\text{cycle/s}}\right) \quad \text{and} \quad \frac{\tau}{\text{s}} = 2\pi\left(\frac{\text{rad/s}}{q}\right)$$

in accordance with the particular case of a rotating crank or vector completing one cycle per rev or turn. From this we may deduce the numerical identities

$$|\omega| = 2\pi\,|n| = 2\pi\,|f| \quad \text{and} \quad |\tau| = \frac{2\pi}{|\omega|} = \frac{1}{|f|}$$

where $|\omega|$, $|f|$ and $|\tau|$ represent numbers only. We usually see these familiar equations written (perhaps unfortunately without the numeric indicator | |) as

$$\omega = 2\pi n = 2\pi f \quad \text{and} \quad \tau = \frac{2\pi}{\omega} = \frac{1}{f}$$

without any qualifying statement that they are numerical formulae necessarily implying the customary ways of measurement, of which, however, only s (the base unit of time) and rad (the supplementary unit for plane angle) are units of the SI; the others are in the 'helpful in practice' category, as is Celsius temperature (°C). Although they are not units of the SI, it is desirable to write down these customary modes of measurement alongside their numerical values and to recognise that the unity multipliers for conversion of units are

$$\left[\frac{\text{Hz s}}{\text{cycle}}\right], \left[\frac{\text{Hz s}}{2\pi\ \text{rad}}\right] \quad \text{and} \quad \left[\frac{\text{Hz s}}{\text{rev}}\right]$$

on the clear understanding that Hz = cycle/s = $2\pi$ rad/s = rev/s (as in Examples 14 and 15).

We may also note that, since numbers and their units arise only when the quantitative or arithmetical stage of a problem has been reached (see Example 14), algebraic analysis of problems involving frequencies can be carried out using one letter symbol (say $\omega$) designating angular velocity of a rotating arm or vector – alternatively called angle rate or angular frequency. This avoids a plethora of '$2\pi$'s and is easily convertible into any of the other ways of specifying 'frequency' deemed appropriate; e.g. the natural frequency of the undamped oscillatory spring-mass system of Example 14 is $\omega_n$ = 20 rad/s, which may alternatively be expressed as

$$\omega_n = 20\,\frac{\text{rad}}{\text{s}}\left[\frac{\text{Hz s}}{2\pi\ \text{rad}}\right] = 3.18\ \text{Hz or } 3.18\ \text{cycle/s} = f_n$$

and the periodic time as

$$\tau = \frac{2\pi\ \text{rad}}{\omega} = \frac{\pi}{10}\ \text{s}$$

or $$\tau = \frac{\text{cycle}}{f_n} = \frac{\text{cycle}}{3.18\ \text{Hz}} \left[\frac{\text{Hz s}}{\text{cycle}}\right] = 0.314\ \text{s}$$

Alternatively, $$|\tau| = \frac{2\pi}{|\omega|} \quad \text{or} \quad \frac{1}{|f|}$$

$$= \frac{2\pi}{20} \quad \text{or} \quad \frac{1}{3.18}$$

i.e. $$\tau = |\tau|\,\text{s} = 0.314\ \text{s}$$

# Appendix 5

## Amount of substance

'Mole' (unit symbol mol) is the name of a unit used mainly by chemists and physicists when expressing 'amount of substance' in quantitative analysis. The number (called Avogadro's number $B_m = 6.02252 \times 10^{23}$) of designated particles (molecules, atoms, ions or electrons, etc.) which, by definition, constitute one mole, leads to

$$\left[\frac{0.012 \text{ kg}}{B_m \times m\,(^{12}\text{C})}\right] = 1 \text{ (unity)},$$

where $m\,(^{12}\text{C})$ is the mass of an atom of carbon-12.

Alternatively,

$$\left[\frac{1 \text{ g}}{B_m \times m_a}\right] = 1$$

where $m_a$ is the (unified) atomic mass unit defined as 1/12 the mass of one atom of carbon-12.

The 'amount of substance' ($n$) is equal to the mass ($m$) of the substance divided by the mass of the specified entity per mole (called molar mass, $M$), i.e. $n = m/M$. Thus, in this sense 'amount of substance' is not the same as mass but is proportional to the number, $B$, of the specified particle in the substance, and may be written

$$n = \frac{B}{B_m} \text{ mol} = \frac{B}{L}$$

where $L = B_m/\text{mol} = 6.02252 \times 10^{23}/\text{mol}$ (called Avogadro's constant, which is the same for all substances and equal to Avogadro's number ($B_m$) per mole).

In the case of pure substances composed of identical molecules (e.g. $O_2$, $CO_2$, $H_2O$, $NH_3$, $C_7H_{16}$, etc.) the relative molecular mass,

$M_r$ (formerly called 'molecular weight'),of substance Z of mass per molecule $m$(Z) is

$$M_r(\mathrm{Z}) = \frac{m(\mathrm{Z})}{m_a} = \frac{m(\mathrm{Z}) \times 12}{m(^{12}\mathrm{C})}$$

and the molar mass of substance Z (i.e. of a number, $B_m$, of molecules of Z) is

$$M(\mathrm{Z}) = B_m \times m(\mathrm{Z}) = \frac{m(\mathrm{Z})}{m(^{12}\mathrm{C})} \times 0.012 \text{ kg or } M_r(\mathrm{Z}) \text{ g per mol}$$

For example, for oxygen $M_r(O_2) = 2 \times 16 = 32$ and the molar mass is $M(O_2) = 0.032$ kg/mol; similarly, $M(C_7H_{16}) = (7 \times 12 + 1 \times 16)$ g/mol $= 0.10$ kg/mol.

Thus, in general (*with molecules as the specified entities*), the molar mass of a substance is $M = M_r$ g/mol, from which we may deduce that

$$1 \text{ mol} = \frac{M_r}{M/\mathrm{g}}$$

This may be translated into words as: one mole is that amount of substance whose molar mass in grams (i.e. $M$/g) is equal to its relative molecular mass ($M_r$).

Hence the 'amount of substance' = total mass/molar mass, i.e.

$$n = \frac{m}{M} = \frac{m}{M_r \text{ g/mol}} = \left(\frac{m/\mathrm{g}}{M_r}\right) \text{mol}$$

from which we see that

$$\left(\frac{n}{\mathrm{mol}}\right) = \left(\frac{10^3}{M_r}\right) \left(\frac{m}{\mathrm{kg}}\right)$$

Consequently we may state that the number of moles of substance is proportional to the mass; and if the mass is measured in kilograms, the constant of proportionality is $(10^3/M_r)$, where $M_r$ is the relative molecular mass of the substance. For example, for $H_2O$, $M_r = 18$; hence 18g of $H_2O$ corresponds to an amount of substance

$$n = \left(\frac{10^3}{18}\right) \left(\frac{18\mathrm{g}}{\mathrm{kg}}\right) \text{mol} = 1 \text{ mol}$$

*The mole applied to gases*

Since, for an ideal gas, $PV = RmT = \bar{R}nT$, the volume $V = (\bar{R}T/P)\,n = (\bar{R}T/PL)B$. Hence, at constant temperature ($T$) and pressure ($P$), the volume $V \propto n \propto B$, from which we may make the following deductions:

(a) Avogadro's law, namely *equal volumes of different gases at the same temperature and pressure contain the same number of molecules,* for, since $n = B/L = B/B_m$ mol, then $B = (n/\text{mol})\,B_m$ and $B = B_m = 6.02252 \times 10^{23}$ if $n = 1$ mol.

Also, *at constant temperature and pressure the volume occupied by one molecule of different gases is the same,* namely $V/B = \bar{R}T/PL$.

Thus it follows that the volume of 1 mol of gas (e.g. 0.032 kg of $O_2$; 0.028 kg of $N_2$; 0.029 kg of air, etc.) at 0°C or 273.15 K and 760 mm Hg or 101.325 kN/m$^2$ is the same, namely 22.418 $\times 10^{-3}$ m$^3$ or *22.418 litres.*

(b) Since $V = (\bar{R}T/P)\,n$, the volumetric ($V$) analysis of a gas mixture, when carried out at constant temperature and pressure, is also a mole ($n$) analysis. For example, the volumetric analysis of dry exhaust gases from an engine or boiler flue (by means of, say Orsat's analyser, in which absorption of $CO_2$, $O_2$ and CO takes place at atmospheric temperature and pressure) is also a mole analysis, since the respective volumes absorbed are proportional to their corresponding number of moles, namely $V_1 \propto n_1$, $V_2 \propto n_2$, $V_3 \propto n_3$ (see Example 23 (b)).

In the case of a mixture of, say, three gases,

$$n = n_1 + n_2 + n_3 = \Sigma n \quad \text{or} \quad \frac{m}{M_x} = \frac{m_1}{M_1} + \frac{m_2}{M_2} + \frac{m_3}{M_3}$$

and

$$m = m_1 + m_2 + m_3$$

Hence the (equivalent) molar mass, $M_x$, of a mixture of gases (conveniently regarded as though it were a single gas) is

$$M_x = \frac{m}{n} = \frac{n_1M_1 + n_2M_2 + n_3M_3}{n_1 + n_2 + n_3} = \frac{\Sigma nM}{\Sigma n}$$

Take, as an example, dry air of percentage composition by volume: oxygen $O_2$ 21.00, nitrogen $N_2$ 78.05, argon Ar 0.95. The molar mass is

$$M_{air} = \frac{(21.00 \times 32 + 78.05 \times 28 + 0.95 \times 39.9)}{100 \text{ mol}}\,\text{g}$$

$$= 0.02895 \text{ kg/mol}$$

There are other problems (e.g. estimation of the volume of water produced by the combustion of, say, 1 litre of liquid fuel) that cannot be solved without using mass as the basic quantity (see Example 25(c)). However, in most problems involving combustion, volumetric analysis of gas mixtures, partial pressures, and problems allied to these, the arithmetical work in calculations in which the mol is used as a unit is somewhat less cumbersome than if the unit of mass (kg) is used, chiefly because the molar (or universal) gas constant ($\bar{R} = PV/nT$) is used rather than the several specific gas constants ($R = PV/mT$) of the individual gases composing the gas mixture.

# Index